T0350084

Undergraduate Texts in Mathematics

Editors

S. Axler
F.W. Gehring
K.A. Ribet

Springer

New York
Berlin
Heidelberg
Barcelona
Hong Kong
London
Milan
Paris
Singapore
Tokyo

Undergraduate Texts in Mathematics

(continued after index)

David M. Bressoud

Factorization
and Primality Testing

 Springer

David M. Bressoud
Chair, Mathematics and Computer Science Department, Macalester College, Saint Paul, MN 55105 USA

The author wishes to express his gratitude for permission to reprint material from the following sources:

First line of a sonnet by Edna St. Vincent Millay. From *Collected Papers*, Revised and Expanded Edition, Harper and Row, 1988. Copyright 1923, 1951 by Edna St. Vincent Millay and Norma Millay Ellis. Reprinted by permission.

Table in Sec. 8.6 reprinted from "The Multiple Polynomial Quadratic Sieve," Robert Silverman, *Mathematics of Computation*, (1987), Vol. 48, No. 177, pp. 329–339, by permission of the American Mathematical Society and Robert Silverman.

Group theory defined by James R. Newman. From *The World of Mathematics*, Tempus Books. Copyright 1988 by Ruth G. Newman. Reprinted by permission.

With 2 illustrations.

Mathematics Subject Classification (1991): 11-01, 11-04, 11A51, 11Y05, 11Y11, 11NXX

Library of Congress Cataloging-in-Publication Data
Bressoud, David M., 1950–
    Factorization and primality testing / David M. Bressoud.
      p.   cm.—(Undergraduate texts in mathematics)
    ISBN 0-387-97040-1 (alk. paper)
    1. Factorization (Mathematics)   2. Numbers, Prime. I. Title.
    II. Series.
    QA161.F3B73   1989
    512′.74—dc20   89-19690

Photocomposed from a LaTeX file.
Printed and bound by Braun-Brumfield, Inc., Ann Arbor, MI.
Printed in the United States of America.

9 8 7 6 5 4 3

ISBN 0-387-97040-1 Springer-Verlag New York Berlin Heidelberg
ISBN 3-540-97040-1 Springer-Verlag Berlin Heidelberg New York SPIN 10763448

Dedicated to two men who have shown me how to write, my father,

MARIUS L. BRESSOUD, JR.

and my mathematical father,

EMIL GROSSWALD (1912–1989)

Preface

The question of divisibility is arguably the oldest problem in mathematics. Ancient peoples observed the cycles of nature: the day, the lunar month, and the year, and assumed that each divided evenly into the next. Civilizations as separate as the Egyptians of ten thousand years ago and the Central American Mayans adopted a month of thirty days and a year of twelve months. Even when the inaccuracy of a 360-day year became apparent, they preferred to retain it and add five intercalary days. The number 360 retains its psychological appeal today because it is divisible by many small integers. The technical term for such a number reflects this appeal. It is called a "smooth" number.

At the other extreme are those integers with no smaller divisors other than 1, integers which might be called the indivisibles. The mystic qualities of numbers such as 7 and 13 derive in no small part from the fact that they are indivisibles. The ancient Greeks realized that every integer could be written uniquely as a product of indivisibles larger than 1, what we appropriately call *prime* numbers. To know the decomposition of an integer into a product of primes is to have a complete description of all of its divisors. By the time Euclid wrote his "Elements" in Alexandria, about 300 B.C., the question of divisibility was recognized to consist of two problems: the description or recognition of the prime numbers and the factorization into primes of the non-prime or *composite* numbers.

Euclid knew these problems to be of more than aesthetic interest. They are intimately tied to almost every question involving integers. Among the problems considered by the Greeks that we shall study are the generation of Pythagorean triples, the characterization of "perfect" numbers, and the approximation of square roots by rational numbers.

It is therefore surprising that a subject that is so very old should at the same time be so very new. Factorization and primality testing is a very hot area of current research; yet the research is still at a sufficiently elementary level that most of the important breakthroughs made in the past few years are accessible to the undergraduate mathematics or computer science major. I am not just talking about finding a bigger prime or factoring a larger number; it is the theoretical approach to such problems which is still in its infancy. I hope that the student reading this book will share in the sense of excitement of being on the leading edge of new mathematics.

Why is it that these ancient problems have blossomed in the past twenty years? In several ways the explanation comes from the electronic computer. As a tool, it permits the implementarion of algorithms whose complexity made them unthinkable a generation ago. As the computer evolves, it forces the researcher to rethink the algorithms. In the past few years, memory has become cheap and as we have approached the theoretical limit on processing speed, there has been increasing emphasis on parallel processing. In response to these developments, today's most useful algorithms use large amounts of memory and are amenable to being run in parallel. The computer industry itself is a consumer of these algorithms. They have shown themselves to be extremely well suited to push computers to their limits, to reveal the flaws, to set the benchmarks.

Among the factors creating interest in factorization and primality testing, one cannot omit the advent of the RSA public key cryptosystem. Based on the simple observation that it is immensely easier to multiply two large primes together than it is to factor their product, it has made the research on factorization and primality testing of direct, practical interest to government and business and anyone concerned with secure transmission of information.

There is another reason for studying factorization and primality testing. It is my own reason for writing this book. Few other problems in mathematics draw so richly on the entire history of mathematics. The algorithms of Euclid and Eratosthenes, now well over two thousand years old, are as fresh and useful today as when they were first discovered. We will be picking up contributions from Fermat in the 17th century, Euler in the 18th, Legendre, Gauss, Jacobi, and a host of modern mathematicians and computer scientists.

Chapters 1 and 2 present basic problems and solutions which were discovered by the Greeks of the classical era. Two of the most important algorithms in use today in factorization and primality testing, the Euclidean Algorithm and the Sieve of Eratosthenes, come to us from this period. We will also investigate the Greek problem of finding perfect numbers. In Chapter 3, we move to 17th-century Europe and some simple observations about this problem which were made by Pierre de Fermat, observations that will form the theoretical underpinning for many of our future algorithms.

In Chapters 4 and 5 we leap to the present and look at current factorization techniques that depend on the theory that has been built up to this point. We also study the applications of factorization and primality testing to the construction of codes for transmitting secret information.

With Chapters 6 and 7 we return to garnering an understanding of the integers. It is now the late 18th, early 19th century. We will see how some of the basic knowledge found by Fermat is deepened by Euler, Legendre, Jacobi, and most especially Gauss. This gives us the theory needed for Chapter 8 in which the Quadratic Sieve will be explained. This algorithm, less than a decade old, is the most powerful tool for factorization known today.

In Chapter 9 we return to Gauss for insights that will answer many of the questions posed up to this point. Gauss' contributions will also lead us to one of the most useful of the current primality tests. There is a natural break in the text at the end of Chapter 9. A one-semester course usually ends at this point, with a week or two spent highlighting some of the topics of the last five chapters.

In Chapters 10, 11, and 12 we travel briefly back to the ancient Greeks to pick up another thread, another problem that has engendered a chain of solutions and problems through the centuries, that of finding rational approximations to irrational numbers. Again it is Fermat who provides the crucial insight that moves the problem forward into our modern era. In these chapters, as the theory is developed we shall jump to the present to show how it is used in modern algorithms: the Continued Fraction Algorithm, the $p + 1$ factorization algorithm, and the primality tests based on Lucas sequences.

Finally, Chapters 13 and 14 delve into the most recent body of theory to find application in factorization and primality testing, the theory of elliptic curves. Here we will be drawing on results of Hasse and Weil that are only a few decades old. Very little is actually proved in these chapters. The emphasis is instead on explaining what the results mean and how they are used.

A course such as this should not and in fact cannot be taught except in conjunction with a computer. The patterns that Fermat, Euler, Gauss, and others saw, the patterns they discovered through many hours of tedious calculations, can now be generated in seconds. I strongly recommend that each student do all or most the the computer exercises at the ends of the chapters and participate in the search for structure.

To facilitate programming, I have chosen to present all the algorithms as computer programs in a generic structured language that owes much to the "shorthand Pascal" used by Stanton and White in *Constructive Combinatorics*. It is my hope that anyone with a familiarity with programming can readily translate these algorithms into their preferred language.

The actual programming does present one major obstacle. In this book we are often working with integers of 60 or more digits for which we need to maintain total accuracy. While high precision subroutines can be written, they are cumbersome. In teaching this course, I have used REXX, a little known but highly useful language developed by IBM. It is a modern, structured language that is extremely simple and ideally suited to integer calculations. It can be run on any IBM compatible machine from a PC up to a mainframe and operates with arbitrary precision. I have translated all of the algorithms in this book into REXX programs and will gladly send a copy to anyone who requests it.

I want to say a word about a major omission from this book. Almost nothing is said about the computational complexity of the algorithms it contains. This was intentional. The most interesting complexity questions are extremely difficult. My emphasis in this book is primarily on the theory behind the algorithms: how they arise and why they work. Secondarily it is on the actual implementation of these algorithms. I feel that to also include a discussion of complexity would distract from my purpose. The interested reader can find very good discussions of computational complexity in Hans Riesel's *Prime Numbers and Computer Methods for Factorization* and in the articles by Carl Pomerance referenced at the end of Chapter 5.

A note on notation: For small integers of nine or fewer digits I am following the standard international convention of separating ones from thousands from millions by spaces, thus

$$362\,901\,095$$

Once I get to ten or more digits I switch to blocks of five digits, such as

$$57\,29001\,87243\,88921\,98362$$

This makes it much easier to count the total number of digits.

I want to thank all the people who have had a hand in making this book possible, among them George Andrews who first suggested I put together such a course and then encouraged me to write the text, John Brillhart and Hugh Williams who helped me find my way into the relevant literature, Robert Silverman for his comments on Chapter 8, Raymond Ayoub for helpful suggestions on presentation, Rüdiger Gebauer at Springer-Verlag who served as a sounding board for different approaches to presenting the algorithms, the librarians at Penn State who helped me track down names and dates and introduced me to Poggendorf, and the students who put up with various preliminary drafts of this book and found many of the mistakes for me. I also want to thank the National Science Foundation and the National Security Agency whose summer research grants, respectively numbers DMS 85-21580 and MDA904-88-H-2017, gave me some of the time I needed to write this book.

David M. Bressoud

University Park, Pennsylvania

November, 1988

Contents

1

Unique Factorization and the Euclidean Algorithm

1.1 A theorem of Euclid and some of its consequences

The integers larger than 1 are of two types: the *composite* integers which can be written as a product of two integers larger than 1 and the *prime* integers (or primes) which cannot. This book revolves around two questions: Given a composite integer, how do we find a decomposition into a product of integers larger than 1? How do we recognize a prime integer? Our investigation of these questions begins approximately 300 B.C. in the Greek city of Alexandria in what is today Egypt. There Euclid wrote his great work "Elements", best known for its treatment of geometry but also containing three books on the properties of the integers. In Proposition 30 of Book VII, Euclid states the following.

Theorem 1.1 *If a prime divides the product of two integers then it must divide at least one of those integers.*

Before proving this theorem. I want to discuss some of its implications and point out that, as obvious as it looks, it is not trivial. This was driven home very forcefully in the early 19th century when mathematicians were led to consider more general types of integers, what we will call *extended integers*. As an example, we can take our extended integers to be all numbers of the form $m + n\sqrt{10}$, where m and n are ordinary integers. Such extended integers can be added, subtracted, multiplied, and even divided using the equality

$$\frac{a + b\sqrt{10}}{m + n\sqrt{10}} = \frac{a + b\sqrt{10}}{m + n\sqrt{10}} \times \frac{m - n\sqrt{10}}{m - n\sqrt{10}}$$

$$= \frac{a \times m - 10b \times n + (b \times m - a \times n)\sqrt{10}}{m^2 - 10n^2}.$$

It makes sense to say that $m + n\sqrt{10}$ is a divisor of $a + b\sqrt{10}$ if and only if their ratio is another extended integer. Thus $2 + \sqrt{10}$ is a divisor of $12 + 3\sqrt{10}$ because

$$\frac{12 + 3\sqrt{10}}{2 + \sqrt{10}} = \frac{-6 - 6\sqrt{10}}{-6} = 1 + \sqrt{10}.$$

One can now talk about indivisibles in this sytem of extended integers. In Exercise 1.2 it is shown that the numbers $2, 4 + \sqrt{10}$, and $4 - \sqrt{10}$ are indivisibles. I call them indivisibles and not primes because they do not satisfy Theorem 1.1:

$$(4 + \sqrt{10}) \times (4 - \sqrt{10}) = 6,$$

which is divisible by 2, and yet 2 does not divide either $4 + \sqrt{10}$ or $4 - \sqrt{10}$.

This has been a digression into an apparently esoteric and unrelated area. But we will see these extended integers again in Chapters 10 – 14 where they will come to play an important role in factorization and primality testing.

The first two applications of Theorem 1.1 use the special case where the two integers are equal. If a prime p divides a^2, then it must divide a. Therefore if p divides a^2, then p^2 divides a^2.

Theorem 1.2 *The square root of 2 cannot be expressed as a ratio of two integers.*

Proof. Assume that the square root of 2 can be written as m/n where m and n are integers and the fraction has been reduced, that is to say, m and n have no common divisors larger than 1. Squaring both quantities, we have that

$$2 = \frac{m^2}{n^2}, \quad \text{or equivalently}$$

$$2n^2 = m^2.$$

Thus 2 divides m^2. By the special case of Theorem 1.1 we have just explained, 4 divides m^2. But then 4 divides $2n^2$, which means that 2 must divide n^2. Again invoking Theorem 1.1, 2 divides n. That gives us our contradiction because we have shown that 2 must divide both m and n, and yet m and n have no common factors larger than 1.

Q.E.D.

(Q.E.D. stands for *Quod Erat Demonstrandum*, a Latin translation of the phrase ὅπερ ἔδει δεῖξαι with which Euclid concluded most of his proofs. It means "what was to be proved" and signifies that the proof has been concluded.)

Before giving our next application, we need three definitions.

Definition. The decomposition of a positive integer into primes is called its *factorization*. This will usually be expressed by

$$n = p_1^{a_1} \times p_2^{a_2} \times \cdots \times p_r^{a_r},$$

where $p_1, p_2, \ldots, p_r$ are distinct primes.

Definition. If integers m and n have no common primes in their respective factorizations, we say that m and n are *relatively prime*. Thus $45 = 3 \times 3 \times 5$ and $98 = 2 \times 7 \times 7$ are relatively prime.

Definition. Three integers (x, y, z) which satisfy

$$x^2 + y^2 = z^2,$$

are called a *Pythagorean triple*. If they are all positive and have no common factors we call them a *fundamental triple*. Because of the symmetry in x and y, we will consider (x, y, z) and (y, x, z) to be the same triple.

The best known Pythagorean triples are $(3, 4, 5)$ and $(5, 12, 13)$. The triple $(6, 8, 10)$ is Pythagorean, but it is not a fundamental triple. It is obtained trivially from $(3, 4, 5)$ by multiplying each integer by two. It should be clear that if we want to find all Pythagorean triples, it is sufficient to first find all fundamental triples and then take the multiples of the fundamental triples. The problem of generating all fundamental triples is solved in the following theorem.

Theorem 1.3 *Given any pair of relatively prime integers, say (a, b), such that one of them is odd and the other even and $a > b > 0$, then*

$$(a^2 - b^2, 2ab, a^2 + b^2)$$

is a fundamental triple. Furthermore, every fundamental triple is of this form.

Note that $(3, 4, 5)$ corresponds to $a = 2$, $b = 1$ while $(5, 12, 13)$ comes from $a = 3$, $b = 2$.

Proof. It is left as Exercise 1.6 to verify that the triple in question is indeed a fundamental triple. The more interesting part of this theorem is that there are no other fundamental triples.

Let (x, y, z) be a fundamental triple. Since x, y, and z are not all even, at most one of them is even. If x and y are both odd, then it is easily verified that x^2 and y^2 are each one more than a multiple of 4, and so z^2 must be two more than a multiple of 4. But that says that z^2 is divisible by 2 and not by 4, a contradiction of Theorem 1.1. Thus either x or y is even. By symmetry in x and y we can assume it is y that is even, say $y = 2m$.

We can use the Pythagorean equation to obtain an equation for m:

$$
\begin{aligned}
y^2 &= z^2 - x^2 \\
&= (z - x) \times (z + x), \\
m^2 &= \frac{z - x}{2} \times \frac{z + x}{2}.
\end{aligned}
$$

Since x and z are each odd, both $(z - x)/2$ and $(z + x)/2$ are integers. Furthermore, they are relatively prime because any common divisor would have to divide both their sum (which is z) and their difference (which is x), and x and z have no common divisors.

For each prime p which divides $(z - x)/2$, p divides m^2 and so p^2 divides m^2. Since p does not divide $(z + x)/2$, p^2 must divide $(z - x)/2$. Thus the factorization of $(z - x)/2$ will have only even exponents, which is another way of saying that $(z - x)/2$ is a perfect square. Similarly, $(z + x)/2$ must be a perfect square. Let us write

$$
\begin{aligned}
\frac{z + x}{2} &= a^2, \\
\frac{z - x}{2} &= b^2.
\end{aligned}
$$

As we have just shown, a and b are relatively prime. Since $a^2 + b^2 = z$ is odd, one of a or b is odd and the other even. Also, a is larger than b. If we now solve for x, y, and z we find that:

$$
\begin{aligned}
z &= a^2 + b^2, \\
x &= a^2 - b^2, \\
y &= 2ab.
\end{aligned}
$$

Q.E.D.

1.2   The Fundamental Theorem of Arithmetic

One of the most important consequences of Theorem 1.1 is the following:

Theorem 1.4 (Fundamental Theorem of Arithmetic) *Factorization into primes is unique up to order.*

What this says is that there may be several ways of ordering the primes that go into a factorization:

$$30 \ = \ 2 \times 3 \times 5, \text{or}$$
$$= \ 3 \times 5 \times 2,$$

but we cannot change the primes that go into the factorization. In our extended integers of the form $m + n\sqrt{10}$ this is not true. As an example, 6 has two distinct factorizations into indivisibles:

$$6 \ = \ 2 \times 3$$
$$= \ (4 + \sqrt{10}) \times (4 - \sqrt{10}).$$

Proof. We will actually prove that every integer with non-unique factorization has a proper divisor with non-unique factorization. If there were integers with non-unique factorization, then eventually we would be reduced to a prime with non-unique factorization, and that would contradict the fact that it is a prime and thus has no positive divisors other than 1 and itself.

Let n be an integer with non-unique factorization:

$$n \ = \ p_1 \times p_2 \times \cdots \times p_r$$
$$= \ q_1 \times q_2 \times \cdots \times q_s,$$

where the primes are not necessarily distinct, but where the second factorization is not simply a reordering of the first. The prime q_1 divides n and so it divides the product of the p_i's. By repeated application of Theorem 1.1, there is at least one p_i which is divisible by q_1. If necessary, reorder the p_i's so that q_1 divides p_1. Since p_1 is prime, q_1 must equal p_1. This says that

$$\frac{n}{q_1} \ = \ p_2 \times p_3 \times \cdots \times p_r$$
$$= \ q_2 \times q_3 \times \cdots \times q_s.$$

Since the factorizations of n were distinct, these factorizations of n/q_1 must also be distinct. Therefore n/q_1 is a proper divisor of n with non-unique factorization.

Q.E.D.

Hopefully, this has given you some idea of the importance of Theorem 1.1. To prove it we will need a very powerful lemma, the essence of which is contained in Propositions 1 and 2 of Book VII in Euclid's "Elements".

Definition. Given integers a and b, the *greatest common divisor* of a and b is the largest positive integer which divides both a and b. We will denote this by $gcd(a, b)$.

Lemma 1.5 *Let a and b be integers and let $g = gcd(a, b)$. Then there exist integers m and n such that*

$$g = m \times a + n \times b.$$

For example let $a = 21$ and $b = 6$. Then $g = 3$ and $m = 1$, $n = -3$ will work:

$$3 = 1 \times 21 + (-3) \times 6.$$

Note that there are other values of m and n that will also work:

$$3 = (-1) \times 21 + 4 \times 6.$$

Before proving this lemma, we show that it does imply Theorem 1.1.

Proof of Theorem 1.1. Let p be a prime which divides $a \times b$. If p divides a, then we are done. If not than $gcd(p, a) = 1$ because 1 is the only other positive integer that divides p. By Lemma 1.5 we can find integers m and n such that

$$1 = m \times p + n \times a.$$

Multiplying both sides by b yields

$$b = m \times p \times b + n \times a \times b.$$

Since p divides $a \times b$, it divides both summands on the right-hand side and so divides their sum, which is b. Thus if p does not divide a then it must divide b.

Q.E.D.

1.3   The Euclidean Algorithm

It appears as if we have exchanged the proof of something that is obvious, namely Theorem 1.1, for a proof of a much less obvious statement, Lemma 1.5. In fact, there is a very nice proof of Lemma 1.5 which not only proves the existence of m and n but explicitly shows us how to calculate them and how to calculate the greatest common divisor. This constructive proof is called the *Euclidean Algorithm* and it will constantly recur as an essential subroutine in factorization and primality testing algorithms.

The key to the Euclidean Algorithm is the following basic property of division in the integers:

Given integers a and b, $b \neq 0$, there exist integers m and r such that

$$a = m \times b + r, \quad \text{with} \quad 0 \leq r < |b|.$$

Proof of Lemma 1.5. We will assume that a and b are positive. Using the fact given above, we have that

$$a = m_1 \times b + r_1, \quad 0 \leq r_1 < b.$$

If $r_1 = 0$ then b divides a, b is the greatest common divisor, and we can choose $m = 0$, $n = 1$. If not, then we can divide b by r_1:

$$b = m_2 \times r_1 + r_2, \quad 0 \leq r_2 < r_1.$$

If $r_2 = 0$, we stop here. If not, then we continue, now dividing r_1 by r_2:

$$r_1 = m_3 \times r_2 + r_3, \quad 0 \leq r_3 < r_2.$$

This process is continued until the remainder is 0, which must eventually happen since the remainders are always non-negative and each remainder is strictly smaller than the previous one. We write down the last two equalities:

$$
\begin{aligned}
r_{k-2} &= m_k \times r_{k-1} + r_k, \quad 0 < r_k < r_{k-1}, \\
r_{k-1} &= m_{k+1} \times r_k + 0.
\end{aligned}
$$

The last non-zero remainder, r_k, is the greatest common divisor of a and b. To see this we work back up the list of equalities. By the last equality, r_k divides r_{k-1}. By the second last equality, since it divides r_k and r_{k-1}, it also divides r_{k-2}. ... By the third equality, since it divides r_3 and r_2, it also divides r_1. By the second equality, it also divides b. By the first equality, r_k divides a. Thus r_k is a common divisor of a and b.

To show that r_k is the largest common divisor, let d be any other common divisor. Since d divides both a and b, it must divide r_1 by the first equality. Continuing down the list, we see that d must divide $r_2, r_3, \ldots, r_k$ and therefore d is less than or equal to r_k.

We now use these equations to find the m and n such that

$$r_k = m \times a + n \times b.$$

By the first equation, r_1 can be written as an integer times a plus an integer times b.

$$r_1 = 1 \times a + (-m_1) \times b.$$

By making this substitution for r_1 in the second equality, we can write r_2 as an integer times a plus an integer times b:

$$
\begin{aligned}
r_2 &= b - m_2 \times r_1 \\
&= b - m_2 \times (a - m_1 \times b) \\
&= -m_2 \times a + (1 + m_1 \times m_2) \times b.
\end{aligned}
$$

Continuing down the list of equations, each r_i can be written as an integer times a plus an integer times b, and this proves the lemma.

Q.E.D.

Let us take as an example $a = 1239$ and $b = 168$:

$$
\begin{aligned}
1239 &= 7 \times 168 + 63, \\
168 &= 2 \times 63 + 42, \\
63 &= 1 \times 42 + 21, \\
42 &= 2 \times 21 + 0.
\end{aligned}
$$

$gcd(1239, 168) = 21.$

$$
\begin{aligned}
63 &= 1239 - 7 \times 168, \\
42 &= 168 - 2 \times 63 \\
&= 168 - 2 \times (1239 - 7 \times 168) \\
&= -2 \times 1239 + 15 \times 168, \\
21 &= 63 - 42 \\
&= (1239 - 7 \times 168) - (-2 \times 1239 + 15 \times 168) \\
&= 3 \times 1239 - 22 \times 168 \\
m &= 3; n = -22.
\end{aligned}
$$

The proof of Lemma 1.5 also implies the following useful result.

Theorem 1.6 *If $g = gcd(a, b)$ and if d is any common divisor of a and b, then d divides g.*

1.4   The Euclidean Algorithm in practice

Because this algorithm will be iterated so often in the programs to be written later, it is important to streamline it as much as possible. You should have two separate subroutines, one to use when all you need to find is the greatest common divisor and thus can ignore the m_i's, the other for those less frequent occasions when you need to find m and n as well as the gcd.

In the following algorithm, I use standard shorthand:

$$a \quad \text{MOD} \quad b$$

to denote the remainder when integer a is divided by the integer b. Thus

$$37 \quad \text{MOD} \quad 5 \ = \ 2$$
$$-24 \quad \text{MOD} \quad 7 \ = \ -3.$$

Algorithm 1.7 *This algorithm computes* $gcd(a, b)$ *using the Euclidean Algorithm.*

INITIALIZE: READ a,b

    Input any two integers **a** *and* **b**.

DIVISION_LOOP: WHILE b $\neq$ 0   DO
         temp $\leftarrow$ b
         b $\leftarrow$ a MOD b
         a $\leftarrow$ temp

    Store the value of **b** *and then compute new values for* **a** *and* **b**.

TERMINATE: WRITE |a|

    The greatest common divisor is the absolute value of the last non-zero **b**, *that is to say* |a|.

It should be noted that the Euclidean Algorithm is very forgiving in the sense that it can be abused considerably and still come up with the correct answer. From the way it is set up, it is natural to enter the larger integer first. The reader should verify that if the smaller number is entered first, then the first iteration of DIVISION_LOOP interchanges these initial integers. Also, if one or both of the initial values are negative, then the absolute value of the final value of a is still the greatest common divisor. Even if one or both of the initial values are zero, this algorithm will return the correct *gcd*.

Algorithm 1.8 *This is Donald E. Knuth's algorithm for computing m and n as well as the gcd. In practice, we will only ever need the value of m so that the second coordinates can be suppressed. Also, as Knuth points out, once m and the gcd are known, the value of n is easily computed from the relationship*

$$gcd = m \times a + n \times b.$$

What makes this algorithm work is that u_i, v_i always satisfy:

$$a \times u_1 + b \times u_2 = u_3, \quad and$$
$$a \times v_1 + b \times v_2 = v_3.$$

INITIALIZE: READ a,b
                   $u_1 \leftarrow 1; \ u_2 \leftarrow 0; \ u_3 \leftarrow a$
                   $v_1 \leftarrow 1; \ v_2 \leftarrow 0; \ v_3 \leftarrow a$

   Input any two integers a and b.

DIVISION_LOOP: WHILE $v_3 \neq 0$ DO
                   $q \leftarrow \lfloor u_3/v_3 \rfloor$
                   CALL NEW_VALUES

   Compute the greatest integer less than or equal to $u_3 \ / \ v_3$ and then reset the values of the u_i and v_i.

TERMINATE: WRITE u_1, u_2, u_3

   The greatest common divisor is the last non-zero value of v_3 which is also the current value of u_3. By the equation satisfied by the u's, m is u_1 and n is u_2.

```
NEW_VALUES:        FOR i = 1 to 3 DO
                      temp ← vᵢ
                      vᵢ ← uᵢ - q × vᵢ
                      uᵢ ← temp
                   RETURN
```

Store the old value of v_i and then compute new values of v_i and u_i. Return new values of u_i and v_i to caller.

Note that in Algorithm 1.8 the third coordinate is precisely running through Algorithm 1.7 and thus when the algorithm terminates, u_3 will be the value of the *gcd*.

Because the *gcd* algorithm is iterated so frequently in most factorization procedures, we want to make it as efficient as possible. What makes the Euclidean Algorithm work is the fact that

$$gcd(a, b) = gcd(b, a - m \times b),$$

for any integer m (see Exercise 1.14). In the Euclidean Algorithm we choose for m that integer which yields the smallest positive value for $a - mb$. We then iterate until one of the integers is zero, at which point the *gcd* is simply the remaining non-zero integer. The optimal value of m is computed by division, and division is a relatively time-consuming operation. There are several suggestions for speeding up the computation of the *gcd* by choosing less than optimal values for m.

One of the most practical alternatives is the binary *gcd* algorithm proposed by Josef Stein in 1961. It takes $m = 1$ and makes use of the fact that division by 2 is extremely fast, especially in machine language. If a and b are both even then

$$gcd(a, b) = 2 \times gcd(a/2, b/2),$$

while if a is odd and b is even then

$$gcd(a, b) = gcd(a, b/2).$$

Algorithm 1.9 *A binary gcd algorithm.*

```
INITIALIZE:        READ a,b
                   e ← 0
```

Input positive integers a and b. e counts the power of 2 in the gcd.

```
PULL_TWOS:              WHILE a and b are even DO
                        a ← a/2
                        b ← b/2
                        e ← e + 1
                        a ← REDUCE(a)
                        b ← REDUCE(b)


SUBTRACTION_LOOP:       WHILE b ≠ 0 or 1 DO
                        c ← |a - b|
                        a ← MINIMUM(a,b)
                        b ← REDUCE(c)


TERMINATE:              IF b = 0 THEN gcd = 2ᵉ×a
                        IF b = 1 THEN gcd = 2ᵉ.
                        WRITE gcd


REDUCE(x):              WHILE x is even DO
                        x ← x/2
                        RETURN
```

Pull all factors of 2 out of x. Return new value of x.

The reader is asked to compare running times on Algorithms 1.7 and 1.9. In a high level language, Algorithm 1.7 is usually faster. Algorithm 1.9 is most efficient when written directly in an assembly language.

1.5   Continued fractions, a first glance

We conclude this chapter with a curious phenomenon that comes out of the Euclidean Algorithm and which will play a very important role later in the development of factorization techniques and primality tests.

Theorem 1.10 *Let a and b be integers and let the Euclidean Algorithm run as follows:*

$$a = m_1 \times b + r_1$$
$$b = m_2 \times r_1 + r_2$$
$$\cdot$$
$$\cdot$$
$$\cdot$$
$$r_{k-1} = m_{k+1} \times r_k + 0.$$

The fraction a/b can then be expressed as

$$\frac{a}{b} = m_1 + \cfrac{1}{m_2 + \cfrac{1}{m_3 + \cdots \cfrac{1}{m_{k+1}}}}.$$

As an example, if we return to the application of the Euclidean Algorithm to the values $a = 1239$, $b = 168$, then this proposition says that

$$\begin{aligned} \frac{1239}{168} &= 7 + \cfrac{1}{2 + \cfrac{1}{1 + \cfrac{1}{2}}} \\ &= 7 + \cfrac{1}{2 + \cfrac{1}{3/2}} \\ &= 7 + \cfrac{1}{8/3} \\ &= \frac{59}{8}. \end{aligned}$$

A fraction of the form given in Theorem 1.10, where the numerator is one and the denominator is a non-negative integer plus a fraction of the same form is called a *continued fraction*. It follows from this theorem that every rational number can be written as a continued fraction. We leave the proof of this theorem as an exercise.

REFERENCES

T. L. Heath, *The Thirteen Books of Euclid's Elements*, Dover Publ. Co., New York, 1956.

Donald E. Knuth, *The Art of Computer Programming, Vol. 2, Seminumerical Algorithms*, 2nd ed., Addison-Wesley, Reading, MA, 1981.

Josef Stein, Computational Problems Associated with Racah Algebra, *J. Computational Phys.*, 1(1967), 397-405.

1.6 EXERCISES

1.1 Show that $3 + \sqrt{10}$ is a divisor of every extended integer of the form $m + n\sqrt{10}$. An extended integer which divides every extended integer is called a *unit*. All other extended integers in this system are called *non-units*. (For the ordinary integers, 1 and -1 are the only units.) The correct definition of an indivisible in this system of extended integers is an extended integer which cannot be written as the product of two non-units.

1.2 Prove that $2, 3, 4+\sqrt{10}$, and $4-\sqrt{10}$ really are indivisibles in the system of extended integers of the form $m + n\sqrt{10}$. *Hint:* If

$$
\begin{aligned}
m + n\sqrt{10} &= (a + b\sqrt{10}) \times (c + d\sqrt{10}), \text{ then} \\
m - n\sqrt{10} &= (a - b\sqrt{10}) \times (c - d\sqrt{10}), \text{ and so} \\
m^2 - 10n^2 &= (a^2 - 10b^2) \times (c^2 - 10d^2).
\end{aligned}
$$

If $a + b\sqrt{10}$ and $c + d\sqrt{10}$ are not units, then in each of these four cases

$$a^2 - 10b^2 \text{ and } c^2 - 10d^2$$

must be $2, -2, 3$, or -3. Show that any perfect square is a multiple of 5 or 1 more or less than a multiple of 5, and therefore

$$a^2 - 10b^2 = 2, -2, 3, \text{ or } -3,$$

has no integral solutions.

1.3 Prove that the square roots of 3 and 5 cannot be written as rational numbers.

1.4 Prove that if n is a positive integer which is not the square of another integer than the square root of n cannot be written as a rational number.

1.5 Find all fundamental Pythagorean triples (x, y, z) with x and y less than 50.

1.6 Prove that if a and b are relatively prime, $a > b > 0$, and one is odd, the other even, then

$$(a^2 - b^2, 2ab, a^2 + b^2)$$

is a fundamental triple.

1.7 Show that if x, y, and z are positive integers which satisfy

$$x^2 + 2y^2 = z^2,$$

and they have no common divisor, then there exist relatively prime integers a and b where b is odd such that

$$
\begin{aligned}
x &= |2a^2 - b^2|, \\
y &= 2ab, \text{ and} \\
z &= 2a^2 + b^2.
\end{aligned}
$$

1.8 To say that d divides a means that there is an integer m such that $a = d \times m$. Prove that the Fundamental Theorem of Arithmetic implies Theorem 1.1.

1.9 Using the Euclidean Algorithm and hand calculation, find

$$gcd(31408, 2718).$$

1.10 Let $lcm(a, b)$ denote the least common multiple of a and b. Prove that $lcm(a, b) = (a \times b)/gcd(a, b)$.

1.11 Write a program to implement Algorithm 1.7 and test it on the pairs

$$(31\,408, 2718),$$

$$(21\,377\,104, 12\,673\,234),$$

$$(355\,876\,536, 319\,256\,544),$$

$$(84187\,85375, 78499\,11069),$$

1.12 Show that if the absolute value of a is less than the absolute value of b, then the first iteration of DIVISION_LOOP in Algorithm 1.7 reverses the order of these values.

1.13 Analyze what happens in Algorithm 1.7 if a and/or b is negative. (*Note*: Replacing v by $-v$ does not change the value of u MOD v.)

1.14 Prove that if m is an integer, then the set of common divisors of a and b is the same as the set of common divisors of b and $a - mb$, and thus

$$gcd(a, b) = gcd(b, a - m \times b).$$

1.15 Write a program to implement Algorithm 1.8 and test it on the pairs of values that appear in exercise 1.11.

1.16 In Algorithm 1.8, verify that after each iteration of DIVISION_LOOP the equations

$$
\begin{aligned}
u_3 &= u_1 \times a + u_2 \times b, \\
v_3 &= v_1 \times a + v_2 \times b,
\end{aligned}
$$

are still satisfied.

1.17 What are the values of v_1 and v_2 when Algorithm 1.8 terminates?

1.18 Find two other integral solutions to the equation

$$1239 \times m + 168 \times n = 21.$$

1.19 Show that there are infinitely many integral solutions of

$$1239 \times m + 168 \times n = 21.$$

Describe how they are generated.

1.20 Write a program to implement Algorithm 1.9 and test it on the pairs of values that appear in Exercise 1.11. Compare running times with Algorithm 1.7.

1.21 Prove that Algorithm 1.9 will eventually terminate.

1.22 Prove Theorem 1.10. *Hint:* Start by rewriting the successive equalities of the Euclidean Algorithm as

$$
\begin{aligned}
a/b &= m_1 + r_1/b \\
b/r_1 &= m_2 + r_2/r_1 \\
r_1/r_2 &= m + 3 + r_3/r_2 \\
&\quad \cdot \\
&\quad \cdot \\
&\quad \cdot \\
r_{k-2}/r_{k-1} &= m_k + r_k/r_{k-1} \\
r_{k-1}/r_k &= m_{k+1}.
\end{aligned}
$$

2

Primes and Perfect Numbers

> "It is recorded that all God's works were completed in six days, because six is a perfect number. ... For this is the first number made up of divisors, a sixth, a third, and a half, respectively, one, two, and three, totaling six."
>
> - St. Augustine of Hippo (The City of God)

2.1 The Number of Primes

With this chapter we begin the process of finding the primes and factoring the composite integers. The first question that arises is whether or not the list of primes is finite. If it were then we could, at least in theory, publish a book containing all the prime numbers and anyone wanting to determine whether an integer were prime would only have to look it up. Unfortunately, there is no limit to the number of primes, a fact which was known to Euclid.

Theorem 2.1 *There are infinitely many primes.*

Proof. Let us assume that there are only finitely many primes, then we can list them all:

$$p_1, p_2, \ldots, p_r.$$

Let P be their product, a very big number but still finite:

$$P = p_1 \times p_2 \times \ldots \times p_r.$$

We now consider $P + 1$ which is an integer and so can be factored into primes. But since all of the primes divide P, none of them divide $P + 1$, since if p_i divides P and it divides $P + 1$, then it must divide 1. This is our contradiction.

Q.E.D.

Observe that all that this proof does for us is prove that there are in-finitely many primes. It is useless in trying to generate the primes. If we know the first n primes, this will give us a new prime, but probably not the next prime. Also, this does *not* promise that $P + 1$ will be a prime. For example:

$$(2 \times 3 \times 5 \times 7 \times 11 \times 13) + 1 = 30031$$
$$= 59 \times 509.$$

All it promises is that $P + 1$ will not have any prime factors from our list.

The next question to arise naturally is how thickly are the primes spread among the integers. A famous result conjectured by several mathematicians at the end of the 18th century and not proved until the end of the 19th century by Jacques Hadamard (1865-1963) and Charles-Jean de la Vallée-Poussin (1866-1962) is the following theorem which I shall not prove in this book.

Theorem 2.2 *The number of primes less than or equal to n is asymptotically $n/(\log n)$. That is to say, if $p(n)$ denotes the number of primes less than or equal to n, then*

$$\frac{p(n)}{n/(\log n)}$$

approaches 1 in the limit as n approaches infinity.

This estimate is surprisingly accurate considering its simplicity. I give below some sample values:

n	$n/(\log n)$	actual number of primes
10^3	$144.7\ldots$	168
10^6	$72\,382.4\ldots$	78\,498
10^9	$48\,254\,942.4\ldots$	50\,847\,478

What this tells us is that there are a lot of primes. They are a lot more frequent than perfect squares, for example. This is good news if we want to find big primes because it tells us that our chances of hitting one just by randomly choosing big numbers are pretty good.

As an example, let us assume that we want to find a prime with 100 digits, something just a little smaller than 10^{100}. In Chapter 4 we will see a situation in which we want a hundred digit prime. The ratio of hundred digit primes to hundred digit integers is approximately:

$$\frac{1}{\log(10^{100})}, \text{ or about } \frac{1}{230}.$$

We can greatly increase our odds just by ignoring multiples of 2 and 3 which make up two-thirds of the hundred digit integers. The odds of choosing a prime at random are improved to approximately 1 in 77.

If we start with a hundred digit integer which is one more than a multiple of 6 and then alternately add 4 and then 2 to the last number generated until we have a list of 100 numbers, the odds are very good (about 73%) that we have at least one prime in our list. If we get our computer to generate a thousand integers in this way, we are virtually assured of having found a prime (99.9998% confidence). There is only one small problem: How do we know which numbers in our list are prime?

In this chapter, we will be looking at two extremely old algorithms for finding primes and factoring. While they will not help us much with our hundred digit integers, they contain ideas that eventually will.

Incidentally, the current state of the art is that "generic" integers of up to about 100 digits can be factored. Much larger numbers can be recognized as primes, however. Up to 300 digits is considered computationally feasible.

2.2   The Sieve of Eratosthenes

The first algorithm is called the Sieve of Eratosthenes and is attributed to this Greek mathematician from Cyrene in what is now Libya who lived about 276-194 B.C. and taught in Alexandria. To find all primes less than or equal to n, we list all the integers from 2 to n. We then work our way down the list. The first integer (namely 2) must be prime. We cross off all multiples of 2 which are larger than 2. The first integer after 2 which has not been crossed off (namely 3) must be prime. We cross off all multiples of 3 which are larger than 3. We continue in this manner. When we have found a new prime, we cross off all multiples of that new prime which are larger than the prime itself and then move to the next integer which has not been crossed off and which must again be prime.

One of the worksavers that Eratosthenes realized was that we do not have to continue this all the way up to n. Once we have found a prime larger than the square root of n, all of the remaining integers which have not been crossed off must be prime. If any of them were composite then they would have to have a factor less than or equal to their square root. If $n = a \times b$ then $a \le \sqrt{n}$ or $b \le \sqrt{n}$.

Algorithm 2.3 *The sieve of Eratosthenes to find all primes less than or* *equal to a given integer n.*

```
INITIALIZE:    READ n
               FOR i = 2 to n DO
                     aᵢ ← i
               j ← 2.
```

n is the upper bound on the list of primes to be generated. The second and third lines set up the values over which we sieve.

```
NEXT_PRIME:    WHILE j² ≤ n DO
                     IF aⱼ ≠ 0 THEN CALL SIEVE(j)
                     j ← j + 1
```

If $a_j \neq 0$ then j is prime.

```
TERMINATE:     FOR i = 2 to n DO
                     IF aᵢ ≠ 0 THEN WRITE i
```

```
SIEVE(i):      t ← 2 × j
               WHILE i ≤ n DO
                     aᵢ ← 0
                     i ← i + 1
               RETURN
```

Proper multiples of j are "crossed out" by resetting a_{t_j} to 0. New values of the a_i are then returned to the caller.

This algorithm has some serious flaws. If n is very large it requires a *lot* of memory. And if you want to use if to prove that n is prime, it would take approximately the square root of n cycles. It does possess, however, a great strength which will come into play in the most powerful known factorization algorithm, the Quadratic Sieve. This strength is that it requires *no* division and essentially no multiplication.

2.3   Trial Division

If we are given an integer less than a million, we can find its prime factors fairly quickly just by using the fact that if it is not a prime, then it must have a factor less than its square root. In this case that would mean less than a thousand. Thus all we need to do is to take a list of all primes less than a thousand and try dividing them into the number to be factored. If none of them divide evenly, then the original number was prime. Each time we find a prime divisor, we divide it out. Once the unfactored portion that remains is less than the square of the last prime we tested, we know that the unfactored portion has to be prime.

There are 168 primes less than or equal to a thousand, and 669 primes less than or equal to five thousand. They can be generated by Algorithm 2.3 and then stored in permanent memory for future use, but you can greatly simplify your memory requirements without sacrificing very much running time by trial dividing by 2, 3, and then all integers up to 1000 or 5000 which are not divisible by 2 or 3.

Up to one thousand, this means 334 trial divisions instead of 168. Up to five thousand, it means 1668 instead of 669. This does slow you down, but by a factor of less than 3 on what is a very speedy calculation. And in practice, when we are confronted with a number to be factored we will only use trial division up to 5000. If the smallest prime factor is larger than 5000, then it is probably quite a bit larger than 5000 and there are faster ways to find it.

Algorithm 2.4 *Factorization by trial division up to a specified maximum* $(= max)$. *Final form will be*

$$n = p_1^{e_1} \times p_2^{e_2} \times \cdots \times p_r^{e_r} \times f,$$

where f is the unfactored portion, f is strictly larger than the square of the largest trial divisor or $f = 1$.

```
INITIALIZE:        READ n, max
                   i ← 0
                   f ← n
```

        i *counts the number of distinct prime factors.*
        f *records the still unfactored portion.*

```
TRY_2&3:           FOR d = 2 to 3 DO
                        IF f MOD d = 0 THEN CALL DIVIDE(f,d,i)
                   d ← 5
                   add ← 2.
```

```
TRY_LOOP:        WHILE d ≤ max and d² ≤ f DO
                     IF f MOD d = 0 THEN CALL DIVIDE(f,d,i)
                     d ← d + add
                     add ← 6 - add

BIG_PRIME:       IF d² > f THEN DO
                     i ← i + 1
                     pᵢ ← f
                     eᵢ ← 1
                     f ← 1
```

If $d^2 > f$, then f *is prime.*

```
TERMINATE:       r ← i
                 FOR i = 1 to r DO
                     WRITE pᵢ, eᵢ
                 WRITE f

DIVIDE(f,d,i):   i ← i + 1
                 pᵢ ← d
                 eᵢ ← 1
                 f ← f/d
                 WHILE f MOD d = 0 DO
                     eᵢ ← eᵢ + 1
                     f ← f/d
                 RETURN
```

When this procedure is called, it means that d *is a prime divisor of* f. *It finds* e_i, *the largest power of* d *that divides* n, *and then returns the new values of* f *and* i *to the caller.*

2.4   Perfect Numbers

Algorithms 2.3 and 2.4 are very simple but slow and inefficient for large numbers. In order to speed things up, we are going to have to delve into some of the patterns that are exhibited by the integers. We are going to have to study the Theory of Numbers. Most of what we will need got its start in some rather esoteric-looking questions posed by the ancient Greeks, and Euclid in particular.

Definition: A positive integer is said to be *perfect* if it is the sum of its proper divisors (those positive divisors strictly less than itself).

The first four perfect numbers are

$$
\begin{aligned}
6 &= 1 + 2 + 3, \\
28 &= 1 + 2 + 4 + 7 + 14, \\
496 &= 1 + 24 + 8 + 16 + 31 + 62 + 124 + 248, \\
8128 &= 1 + 2 + 4 + 8 + 16 + 32 + 64 + 127 + 254 + \\
&\quad + \ 508 + 1016 + 2032 + 4064.
\end{aligned}
$$

Several questions suggest themselves on looking over this admittedly skimpy list of perfect numbers: Are there infinitely many? Are there any odd ones? Is there any simple way of generating them? Do the even perfect numbers have to end in 6 or 8?

The answers to the first two questions are unknown, but it is believed by many that the answer to the first question is "yes" and to the second "no". (It has been shown by Peter Hagis that any odd perfect number must be at least 10^{50}.) We will postpone the answer to the last question until the next chapter. If we restrict ourselves to the even perfect numbers, then there is a relatively simple way of finding them. Consider the factorizations of the first four perfect numbers:

$$
\begin{aligned}
6 &= 2 \times 3, \\
28 &= 2^2 \times 7 = 4 \times 7, \\
496 &= 2^4 \times 31 = 16 \times 31, \\
8128 &= 2^6 \times 127 = 64 \times 127.
\end{aligned}
$$

Definition: Let $M(n) = 2^n - 1$. A *Mersenne prime* is a prime of the form $M(n)$ for some integer n. Thus 3, 7, 31, and 127 are the first four Mersenne primes.

Fr. Marin Mersenne (1588-1648) was among the mathematicians of the early 17th century who worked on the problem of perfect numbers. The special role of these primes had actually been known to Euclid.

Theorem 2.5 *If $M(n)$ is a Mersenne prime, then*

$$
m = 2^{n-1} \times M(n)
$$

is a perfect number.

The proof of this theorem and much of the rest of the work on perfect numbers relies on the following lemma.

Lemma 2.6 $x^k - 1 = (x - 1) \times (1 + x + x^2 + \cdots + x^{k-1})$.

The proof of this lemma is left as an exercise. Note that when $x = 2$, it says that

$$2^k - 1 = 1 + 2 + 4 + \cdots + 2^{k-1}.$$

Proof of Theorem 2.5: If $M(n)$ is prime, then the proper divisors of

$$m = 2^{n-1} \times M(n)$$

are $1, 2, 4, \ldots, 2^{n-1}, M(n), 2 \times M(n), 4 \times M(n), \ldots, 2^{n-2} \times M(n)$. The sum of the proper divisors of m is thus:

$$(1 + 2 + 4 + \cdots + 2^{n-1}) + (1 + 2 + 4 + \cdots + 2^{n-2}) \times M(n)$$
$$= (2^n - 1) + (2^{n-1} - 1) \times (2^n - 1)$$
$$= 2^{n-1} \times (2^n - 1) = m.$$

Q.E.D.

Theorem 2.5 implies that for every Mersenne prime there is an even perfect number. The next theorem states that there are no other even perfect numbers.

Theorem 2.7 *If m is an even perfect number then there is an integer n such that*

$$m = 2^{n-1} \times (2^n - 1),$$

and $2^n - 1$ is prime.

Proof. Write m as

$$m = 2^a \times t,$$

where t is odd and a is at least one (because m is even). Let S be the sum of all the divisors of t. In other words, S is the sum of the odd divisors of m.

We take the divisors of m and split them up into the odd divisors, then those divisors with one factor of 2, then those with two factors of 2, and so on until we get those divisors with a factors of 2. The sum of *all* divisors of m thus looks like:

$$S + 2 \times S + 4 \times S + \cdots + 2^a \times S.$$

This sum includes the divisor m itself, and so if we want the sum of the proper divisors, we must subtract m from the expression given above. Since m is perfect, the sum of the proper divisors is equal to m. We get the following equality:

$$
\begin{aligned}
m &= S + 2 \times S + 4 \times S + \cdots + 2^a \times S - m, \\
&= (2^{a+1} - 1) \times S - m.
\end{aligned}
$$

If we solve this equality for S, we see that:

$$S = \frac{2m}{2^{a+1} - 1}.$$

We now rewrite this last equation using the representation which we have for m:

$$
\begin{aligned}
S &= \frac{2^{a+1} \times t}{2^{a+1} - 1} = \frac{(2^{a+1} - 1 + 1) \times t}{2^{a+1} - 1} \\
&= t + \frac{t}{2^{a+1} - 1} = t + \frac{t}{M(a+1)}.
\end{aligned}
$$

Since S is an integer, $t/M(a+1)$ must also be an integer and thus a divisor of t. Now S is the sum of *all* divisors of t, but we have it written here as a sum of exactly two divisors of t. That means that t has exactly two divisors and so t is prime. The two divisors of a prime are itself and 1 and so:

$$\frac{t}{M(a+1)} = 1,$$

and the theorem is proved.

Q.E.D.

2.5 Mersenne Primes

If we ignore the possibility of odd perfect numbers, then we can characterize the perfect numbers by finding the Mersenne primes. When is $M(n)$ prime? One quick result tells us when it is not prime.

Theorem 2.8 *If n is composite, then $M(n)$ is composite.*

Proof: Let $n = a \times b$ where a and b are each larger than 1. Using Lemma 2.6 we have that

$$
\begin{aligned}
M(n) &= 2^{a \times b} - 1, \\
&= (2^a)^b - 1, \\
&= (2^a - 1) \times (1 + 2^a + 2^{2a} + \cdots + 2^{(b-1) \times a}).
\end{aligned}
$$

Since each of these factors is larger than 1, $M(n)$ is composite.

Q.E.D.

Our problem has been reduced to deciding when $M(p)$ is prime. It starts out looking like it is always prime:

$$
\begin{aligned}
M(2) &= 3, \text{ prime} \\
M(3) &= 7, \text{ prime} \\
M(5) &= 31, \text{ prime} \\
M(7) &= 127, \text{ prime.}
\end{aligned}
$$

But, unfortunately, this does not last:

$$
\begin{aligned}
M(11) &= 2047 = 23 \times 89, \\
M(13) &= 8191, \text{ prime} \\
M(17) &= 131\,071, \text{ prime} \\
M(19) &= 524\,287, \text{ prime} \\
M(23) &= 8\,388\,607 = 47 \times 178\,481.
\end{aligned}
$$

In fact, primes p for which $M(p)$ is prime start to get scarce at this point. The next 13 Mersenne primes have the following values for p:

$$31, 61, 89, 107, 127, 521, 607, 1279, 2203, 2281, 3217, 4253, 4423.$$

$M(4423)$ is a big number. It has 1332 digits. How could anyone possibly know that it is a prime? The answer to this is that there is an extremely fast and simple algorithm known as the Lucas-Lehmer algorithm for testing whether a Mersenne number is a prime. The theory behind it was developed by Edouard Lucas (1842-1891) and it was put into its present simplified form by Derrick H. Lehmer. It only works on Mersenne numbers, and its justification requires some fairly high-powered number theory which we will not get to until section 5 of Chapter 11. But I give it here for your amusement.

Algorithm 2.9 *Test for whether or not* $M(n) = 2^n - 1$ *is prime. It is valid for any odd* $n \geq 3$.

```
INITIALIZE:     READ n
                M ← 2^n - 1
                S ← 4

MYSTERY_LOOP:   FOR i = 2 to n - 1 DO
                    S ← S × S - 2 MOD M.

TERMINATE:      IF S = 0 THEN WRITE M
```

M *is prime if and only if the final value of* S *is* 0.

The limitation here is not the complexity of the algorithm, but the difficulties of doing very high precision arithmetic. For $n = 4423$, we need 2664 digit accuracy. Most multiple precision packages will give you arbitrarily high precision, but the higher the precision, the slower they run.

It should not be surprising that the largest known prime is a Mersenne prime. As of this writing it is

$$2^{216091} - 1,$$

an integer with 65050 digits. It was found by David Slowinski in 1985.

REFERENCES
D. H. Lehmer, "An extended theory of Lucas functions," *Ann. Math.*, **31**(1930), 419-448.

Edouard Lucas, "Théorie des fonctions numériques simplement périodiques," *Amer. J. Math.*, **1**(1878), 184-240, 289-321.

2.6 EXERCISES

2.1 Prove that there are infinitely many primes which are one less than a multiple of four. *Hint*: Show that any integer of the form $4n - 1$ must be divisible by a prime of the same form.

2.2 Using Theorem 2.2, approximately how many hundred digit primes are there? How does this compare with the number of primes with *at most* a hundred digits?

2.3 What is the asymptotic formula for the number of perfect squares less than or equal to n?

2.4 Write a program to implement Algorithm 2.3. Use it to find the primes less than or equal to 5000.

EXERCISES 2.5 – 2.7 USE THE TABLE OF PRIMES GENERATED IN EXERCISE 2.4.

2.5 How many pairs of primes in the table differ by two? A famous unsolved problem asks if there are infinitely many such pairs among all the primes.

2.6 How evenly are the primes in the table divided between those one more than a multiple of four and those which are one less than a multiple of four? Are you prepared to make any conjectures? Can you prove your conjectures?

2.7 Can you find any patterns or unusual clusters in your list of primes?

2.8 Write a program to implement Algorithm 2.4. Use it to factor or prove primality for

$$307\,821, 16\,803\,654, 19\,46852\,76691.$$

2.9 Choose 100 consecutive seven digit numbers and factor them using trial division.

EXERCISES 2.10 – 2.15 USE THE FACTORED NUMBERS FROM EX-ERCISE 2.9.

2.10 How many primes are in your list of factored numbers? How does this compare with the expected number of primes?

2.11 How many perfect squares are in your list of factored numbers? How does this compare with the expected number of perfect squares?

2.12 What is the distribution of the number of distinct primes dividing each of your integers? What is the mean and standard deviation of this distribution?

2.13 What is the distribution of the sizes of the primes dividing your integers? How many of the integers n have a prime factor larger than $n^{3/4}$,

$n^{1/2}$, $n^{1/3}$? Describe the distribution, mean, and standard deviation of the logarithm of the largest prime factor divided by the logarithm of n.

2.14 How many of your 100 integers have the number of distinct prime factors within one standard deviation of the mean AND the logarithm of the largest prime factor divided by the logarithm of n within one standard deviation of the mean?

2.15 How large a number can "usually" be factored using trial division up to 5000? Try your trial division algorithm on 100 consecutive integers of this size and report your results. "Usually" should mean at least 75% of the time and not more than 95% of the time.

2.16 Prove Lemma 2.6 by induction on k.

2.17 In the proof of Theorem 2.7, where did we use the fact that the power of 2 is at least one? That is to say, why can't we use this proof to characterize odd perfect numbers?

2.18 Write a program to implement Algorithm 2.9. Use it to check that $M(89)$ is prime but $M(83)$ is not.

2.19 Use trial division to factor $M(29)$. Can you find any pattern to or properties of the prime divisors of $M(11)$, $M(23)$, and $M(29)$?

2.20 For n less than or equal to 30, when is $2^n + 1$ a prime? Can you make a conjecture on when it will be prime? Try to prove or disprove your conjecture.

3

Fermat, Euler, and Pseudoprimes

"I have found a very great number
of exceedingly beautiful theorems."
– Pierre de Fermat

3.1 Fermat's Observation

We have reduced the problem of finding even perfect numbers to deciding
when $M(p) = 2^p - 1$ is prime. Algorithm 2.9 is a very recent development.
In this chapter we will be starting with some progress made by Pierre de
Fermat (1601-1665) in 1640.

Fermat was working with the list of values of $M(p)$ up to $p = 23$ that
was given in Chapter 2. He observed a curious phenomenon: For each prime
q that divides $M(p)$, the remainder when q is divided by p is 1. If this is
always true, then it greatly simplifies our task of looking for divisors of
$M(p)$ because we can restrict our attention to those primes q which are
one more than a multiple of p.

Observe that there does appear to be something special about primes
going on here because this property does not always hold for $M(n)$ when
n is composite:

$$
\begin{aligned}
M(4) &= 15 = 3 \times 5, \\
M(6) &= 63 = 3 \times 3 \times 7, \\
M(8) &= 255 = 3 \times 5 \times 17.
\end{aligned}
$$

Definition: It is useful to introduce some notation. We will write:

$$a \equiv b \,(\text{mod } m),$$

to mean that m divides $a - b$. Equivalently, if a and b are positive, it means
that

$$a \text{ MOD } m = b \text{ MOD } m.$$

This is read as *a is congruent to b modulo (or mod) m* or *a and b belong
to the same congruence class modulo m*. The smallest non-negative integer

congruent to a modulo m will be called the *residue of a modulo m*. Again, if a is positive, then the residue of a modulo m is a MOD m.

In this notation, Fermat's observation is that if a prime q divides $M(p)$, then

$$q \equiv 1 \ (\text{mod } p).$$

I leave it as an exercise to verify that if

$$a \equiv x \ (\text{mod } m), \quad \text{and} \quad b \equiv y \ (\text{mod } m), \text{then}$$

$$\begin{aligned} a + b &\equiv x + y \ (\text{mod } m), \text{and} \\ a \times b &\equiv x \times y \ (\text{mod } m). \end{aligned}$$

We will also need to use the fact that if $gcd(x, m) = 1$ and if

$$\begin{aligned} a \times x &\equiv b \times x \ (\text{mod } m), \text{then} \\ a &\equiv b \ (\text{mod } m). \end{aligned}$$

The proof of this is also left as an exercise.

Any divisor d of $M(p)$ is a product of primes which divide $M(p)$ and so Fermat's observation implies that any divisor d of $M(p)$ satisfies

$$d \equiv 1 \ (\text{mod } p).$$

In particular, $M(p)$ is a divisor of $M(p)$, that is to say:

$$2^p - 1 \equiv 1 \ (\text{mod } p).$$

If p is not 2, then we can add 1 to each side and then divide by 2 to get the following result.

Theorem 3.1 *If p is an odd prime then*

$$2^{p-1} \equiv 1 \ (\text{mod } p).$$

Note that we do not yet have a proof of this theorem because Fermat's observation is still just an observation and not yet a theorem. Ironically, we will be proving Theorem 3.1 by another method and then using it to prove Fermat's observation.

3.2   Pseudoprimes

For the moment, let us accept Theorem 3.1. It tells us that primes satisfy a very special equality. If we try this equality on composite numbers, we see that it does not seem to work:

$$
\begin{aligned}
2^{4-1} &= 8 &\equiv 0 \,(\mathrm{mod}\ 4), \\
2^{6-1} &= 32 &\equiv 2 \,(\mathrm{mod}\ 6), \\
2^{15-1} &= 16\,384 &\equiv 4 \,(\mathrm{mod}\ 15).
\end{aligned}
$$

This is very nice because we are looking for a way of distinguishing primes from composite numbers. Unfortunately, there are some composite numbers which look like primes from this point of view:

$$2^{341-1} \equiv 1 \,(\mathrm{mod}\ 341), \text{however,}$$

$$341 = 11 \times 31.$$

This motivates the following definition

Definition: If n is odd and composite and n satisfies

$$2^{n-1} \equiv 1 \,(\mathrm{mod}\ n), \tag{3.1}$$

then we say that n is a *pseudoprime*.

Despite the fact that the test we have is not a 100% guarantee of primality, in practice it is very useful. As we will see in the next algorithm, exponentiation can be done very quickly so that most composite numbers are revealed as composite in short order. Furthermore, pseudoprimes are fairly scarce, much scarcer than primes. To give you a feel for their scarcity, there are only 3 pseudoprimes below a thousand:

$$341\ ,\ 561\ ,\ \text{and}\ 645.$$

There are only 245 pseudoprimes below a million (as opposed to 78498 primes). If an integer satisfies Equation (3.1), you know that it is probably a prime and can usually proceed on that basis.

Fermat realized that there was nothing special about the 2 in his theorem and proved the following more general result.

Theorem 3.2 *If p is a prime which does not divide b, then*

$$b^{p-1} \equiv 1 \,(\mathrm{mod}\ p).$$

Definition: If n is an odd composite number which is relatively prime to b and if

$$b^{n-1} \equiv 1 \; (\bmod \; n), \qquad (3.2)$$

then we say that n is a *pseudoprime for the base b*.

We can now strengthen our primality test. If n passes for the base 2, we can also check for bases 3 and 5. For example,

$$3^{341-1} \equiv 56 \; (\bmod \; 341),$$

and we now know that 341 is composite.

Unfortunately, there are composite numbers which are pseudoprimes for all bases to which they are relatively prime. The first such number is

$$561 = 3 \times 11 \times 17.$$

Such numbers are called *Carmichael numbers* after Robert Daniel Carmichael (1879-1967). They are extremely rare, there are only 2163 Carmichael numbers below 25×10^9.

3.3   Fast Exponentiation

The idea behind fast exponentiation is that if the exponent is a power of 2 then we can exponentiate by successively squaring:

$$\begin{aligned} x^8 &= ((x^2)^2)^2, \\ x^{256} &= (((((((x^2)^2)^2)^2)^2)^2)^2)^2. \end{aligned}$$

If the exponent is not a power of 2, then we use its binary representation, which is just a sum of powers of 2:

$$x^{291} = x^{256} \times x^{32} \times x^2 \times x^1.$$

Thus to raise x to the power n requires only about $\log n$ operations.

Warning: Never use an exponentiation symbol (^ or **) to compute exponents when doing factorization or primality testing. It may use logarithms to compute the exponent which will result in some round-off. Even if this is not the case (for example, REXX does use binary exponentiation), you must be careful of overflow. With an accuracy of seventy decimal digits, Algorithm 3.3 given below will quickly compute a^b MOD m for a and m up to 10^{35} and b up to 10^{70}. However, just computing 2^{234} directly would overflow that accuracy.

Algorithm 3.3 *Computes a^b MOD m for $b \geq 0$.*

```
INITIALIZE:    READ a,b,m
               n ← 1


BINARY_LOOP:   WHILE b ≠ 0 DO
                    IF b is odd THEN DO
                        n ← n × a MOD m
                    b ← ⌊b/2⌋
                    a ← a × a MOD m
```

We find the binary representation of b *while at the same time computing successive squares of* a. *The variable* n *records the product of the powers of* a.

```
TERMINATE:     WRITE n
```

The final value of n *is the value of* a^b *MOD* m.

In practice, if you have a large number n that you want to factor or prove prime, you will first use Algorithm 2.4 to check for small divisors. If there are no small divisors, then the next step is to use Algorithm 3.3 to compute

$$2^{n-1} \text{ MOD } n, \quad 3^{n-1} \text{ MOD } n, \quad 5^{n-1} \text{ MOD } n, \quad \text{and } 7^{n-1} \text{ MOD } n.$$

If any of these quantities are not 1, then you know that n is composite. In Chapter 5 we will look at ways of finding large factors.

If all of these are 1, then n is prime or a pseudoprime for the bases 2, 3, 5, and 7, and it is very likely that n really is a prime. We shall call such an n a *probable prime*. The pseudoprime tests are so powerful that the integers which pass them have been dubbed "industrial grade primes" by Henri Cohen. If you really want to prove that n is prime, I will start talking about primality tests in Chapter 9.

3.4   A Theorem of Euler

The easiest way to prove Theorem 3.2 (which includes Theorem 3.1 as a special case) is to prove an even more general result which was found by Leonard Euler (1707-1783) and which will be very important to us. As we have seen, Theorem 3.2 is no longer true if we replace p with a composite integer. Euler found the right result for the composite integers.

Definition: Let $\phi(n)$ denote the number of positive integers less than or equal to n and relatively prime to n. For example:

$$
\begin{aligned}
\phi(4) &= 2, \\
\phi(6) &= 2, \\
\phi(7) &= 6, \\
\phi(15) &= 8.
\end{aligned}
$$

Note that according to this definition, $\phi(1) = 1$.

Theorem 3.4 *Let n and b be positive, relatively prime integers. Then*

$$b^{\phi(n)} \equiv 1 \pmod{n}.$$

If n is prime, then $\phi(n) = n - 1$ and so Theorem 3.2 is a special case of Theorem 3.4.

As an example, consider

$$2^{\phi(15)} = 2^8 = 256 \equiv 1 \pmod{15}.$$

Proof: Let $t = \phi(n)$ and let $a_1, a_2, \ldots, a_t$ be the positive integers less than n which are relatively prime to n. Define $r_1, r_2, \ldots, r_t$ to be the residues of $b \times a_1, b \times a_2, \ldots, b \times a_t \bmod n$. That is to say,

$$b \times a_i \equiv r_i \pmod{n}.$$

We note that if i and j are distinct, then r_i and r_j are also distinct. If they were not then we would have

$$b \times a_i \equiv b \times a_j \pmod{n}.$$

Since $gcd(b, n) = 1$, Exercise 3.2 implies that we can divide by b and thus

$$a_i \equiv a_j \pmod{n},$$

which cannot happen since a_i and a_j are distinct integers between 0 and n.

We also know that each r_i is relatively prime to n because any common divisor of n and r_i would also have to divide a_i. Thus $r_1, r_2, \ldots, r_t$ is a set of $\phi(n)$ distinct integers between 0 and n which are each relatively prime to n. This means that they are exactly the same as $a_1, a_2, \ldots, a_t$, except that they are in a different order. In particular, we have proved that

$$a_1 \times a_2 \times \ldots \times a_t = r_1 \times r_2 \times \ldots \times r_t.$$

Now we use our congruence:

$$r_1 \times r_2 \times \cdots \times r_t \equiv b \times a_1 \times b \times a_2 \times \cdots \times b \times a_t \pmod{n}$$
$$\equiv b^t \times a_1 \times a_2 \times \cdots \times a_t \pmod{n}$$
$$\equiv b^t \times r_1 \times r_2 \times \cdots \times r_t \pmod{n}.$$

We can divide both sides by the product of the r_i's to get

$$1 \equiv b^{\phi(n)} \pmod{n}.$$

Q.E.D.

3.5   Proof of Fermat's Observation

We have finally proved Theorem 3.1. Let us now return to Fermat's original observation that if a prime q divides $M(p)$, then $q \equiv 1 \pmod{p}$. We will need the following lemmas.

Lemma 3.5 *Let x, m, and n be positive integers with m and n relatively prime, then*

$$r = 1 + x + x^2 + \cdots + x^{m-1}, \quad and$$
$$s = 1 + x + x^2 + \cdots + x^{n-1}$$

are relatively prime.

Proof. Let d be any common divisor of r and s. Since r and s are each one more than a multiple of x, d is relatively prime to x. We can assume that m is larger than n. Now d divides

$$r - s = x^n + x^{n+1} + \cdots + x^{m-1}$$
$$= x^n \times (1 + x + \cdots + x^{m-n-1}),$$

and so d is also a divisor of

$$1 + x + x^2 + \cdots + x^{m-n-1}.$$

Now $m - n$ is relatively prime to both m and n, and so we can continue shortening the geometric series that d divides until eventually we will have to have that d divides 1, which means that d is 1.

Q.E.D.

Lemma 3.6 *Let a and b be positive integers, then*

$$gcd(2^a - 1, 2^b - 1) = 2^{gcd(a,b)} - 1.$$

Proof. Let $g = gcd(a, b)$, $a = m \times g$, $b = n \times g$. Then m and n are relatively prime and we have that

$$
\begin{aligned}
2^a - 1 &= (2^g - 1) \times (1 + 2^g + 2^{2 \times g} + \cdots + 2^{(m-1) \times g}), \\
2^b - 1 &= (2^g - 1) \times (1 + 2^g + 2^{2 \times g} + \cdots + 2^{(n-1) \times g}).
\end{aligned}
$$

The proof now follows from Lemma 3.5 with $x = 2^g$.

Q.E.D.

Theorem 3.7 *Let p be a prime and let $M(p) = 2^p - 1$. If d is any divisor of $M(p)$, then*

$$d \equiv 1 \pmod{p}.$$

Proof. As we showed at the beginning of this chapter, it is enough to prove this theorem for prime divisors, say q. We know by Theorem 3.1 that q divides

$$2^{q-1} - 1.$$

But then it has to divide the *gcd* of $2^p - 1$ and $2^{q-1} - 1$, and so by Lemma 3.6, q divides

$$2^{gcd(p,q-1)} - 1.$$

Now the *gcd* of p and $q - 1$ is either 1 or p. If it is 1, then q divides

$$2 - 1 = 1,$$

which cannot happen. That means that p divides $q - 1$ and the theorem is proved.

Q.E.D.

3.6   Implications for Perfect Numbers

The study of perfect numbers led Fermat to his observation and eventually to Euler's theorem (Theorem 3.4). These results will suggest more questions that will lead us on in our study of properties of the integers, and we are ready to bid farewell to perfect numbers until the end of Chapter 11 when the justification of Algorithm 2.9 will drop fortuitously out of our study of continued fractions. But before going, I want to summarize what we have proven about Mersenne primes and answer the question of the last digit in an even perfect number.

The process of testing whether or not $M(p)$ is prime has been considerably simplified by Theorem 3.1. As an example, let us consider

$$M(19) = 524\,287.$$

If it is not a prime, then it must be divisible by a prime congruent to 1 modulo 19 and less than the square root of $M(19)$ which is $724.07\ldots$. Only six primes satisfy these requirements: 191, 229, 419, 571, and 647. None of them divide evenly into $M(19)$ and so $M(19)$ is prime.

The question of the last digit is answered in the next result.

Theorem 3.8 *If n is an even perfect number, then its last digit is either 6 or 8.*

Proof: We know that n looks like

$$n = 2^{p-1} \times (2^p - 1),$$

for some prime p. If $p = 2$, then $n = 6$ and we are okay, so let us assume that p is odd. We want to show that 10 divides either $n - 6$ or $n - 8$.

By Theorem 3.1, we know that

$$2^4 \equiv 1 \pmod 5.$$

Now $p - 1$ is even and so either $p - 1 = 4m$ or $p - 1 = 4m + 2$ for some integer m. We take the first possibility. Then

$$
\begin{aligned}
2^{p-1} &= (2^4)^m \equiv 1 \,(\mathrm{mod}\ 5),\ \text{and} \\
2^p - 1 &= 2^{p-1} \times 2 - 1 \equiv 2 - 1 \equiv 1 \pmod 5,
\end{aligned}
$$

which means that

$$
\begin{aligned}
n &\equiv 1 \times 1 \equiv 1 \pmod 5, \\
n &\equiv 1 \text{ or } 6 \pmod{10}.
\end{aligned}
$$

Since n is even, this case gives us

$$n \equiv 6 \pmod{10}.$$

If $p - 1 = 4 \times m + 2$, then

$$2^{\,p-1} = (2^4)^m \times 4 \equiv 4 \pmod 5, \text{ and}$$
$$2^{\,p} - 1 = 2^{p-1} \times 2 - 1 \equiv 7 \pmod 5, \text{ and so}$$
$$n \equiv 4 \times 7 \equiv 3 \pmod 5,$$
$$n \equiv 3 \text{ or } 8 \pmod{10}.$$

Again, n is even and so in this case we have that

$$n \equiv 8 \pmod{10}.$$

Q.E.D.

Observe that we have proved more than we asked. We now know that if $p = 2$ or $p \equiv 1 \pmod 4$, then $2^{p-1} \times M(p) \equiv 6 \pmod{10}$; if $p \equiv 3 \pmod 4$ then $2^{p-1} \times M(p) \equiv 8 \pmod{10}$.

REFERENCES

R. D. Carmichael, "Note on a New Number Theory Function," *Bull. Am. Math. Soc.*, **16**(1909-1910), 232-238.

R. D. Carmichael, "On composite numbers P which satisfy the Fermat congruence $A^{P-1} \equiv 1 \bmod P$," *Am. Math. Monthly*, **19**(1912), 22-27.

3.7 EXERCISES

IN EXERCISES 3.1 – 3.3, USE THE FACT THAT $a \equiv b \pmod m$ IF AND ONLY IF m DIVIDES $a - b$.

3.1 Prove that if $a \equiv x \pmod m$, and $b \equiv y \pmod m$, then $a + b \equiv x + y \pmod m$ and $a \times b \equiv x \times y \pmod m$.

3.2 Prove that if $gcd(x, m) = 1$ and if $a \times x \equiv b \times x \pmod{m}$, then $a \equiv b \pmod{m}$. Show that the conclusion does not necessarily follow if $gcd(x, m)$ is not 1.

3.3 Let $g = gcd(x, m)$. Prove that $a \times x \equiv b \times x \pmod{m}$ if and only if $a \equiv b \pmod{m/g}$.

3.4 Write a program to implement Algorithm 3.3. Use it to test whether 22123 74139 and 1 97076 83773 are pseudoprimes (base 2).

3.5 Find ten 12-digit probable primes.

3.6 Find a 100-digit probable prime.

3.7 Take the unfactored numbers from your answer to Exercise 2.15 and test whether they are composite or probable primes.

3.8 Prove that if p is prime and $q = 2 \times p + 1$ and q divides $M(p)$, then q must be prime. As an example, verify that 1103 is a prime and that 2207 divides $M(1103)$. By the first part of this exercise, 2207 must be prime.

3.9 Using the idea of Exercise 3.8, show that if p is a prime, q divides $M(p)$, and q is less than p^2, then q must be a prime. As an example, verify that 1153 is a prime and that 267 497 divides $M(1153)$ and so 267 497 must be a prime.

3.10 Exercise 3.9 implies the following primality test: If p is a prime factor of $q - 1$ such that p^2 is larger than q and q divides $M(p)$, then q must be prime. Try using this test on the ten probable primes which you found in Exercise 3.5. For how many of them does it work? When it fails, why does it fail? Does its failure mean that your probable prime is not prime?

3.11 Verify that the following algorithm produces the binary expansion of n:

$$n = b_k \times 2^k + b_{k-1} \times 2^{k-1} + \cdots + b_1 \times 2 + b_0.$$

INITIALIZE: **READ n**
                   i ← -1

Input a non-negative integer n.

```
BINARY_LOOP: WHILE n ≠ 0 DO
                    i ← i + 1
                    bᵢ ← n MOD 2
                    n ← ⌊n/2⌋

TERMINATE:     k ← i
               FOR i = 1 to k DO
                    WRITE bᵢ
```

3.12 The following variation of Algorithm 3.3 is often called peasant multiplication because it has been used for centuries by European peasants who could use it to multiply without needing to memorize multiplication tables. Prove that it does provide the product of a and b as long as b is non-negative, and that it is equivalent to multiplication in base 2.

```
INITIALIZE:    READ a,b
               n ← 0

BINARY_LOOP: WHILE b ≠ 0 DO
                    IF b is odd THEN DO
                         n ← n + a
                    b ← ⌊b/2⌋
                    a ← a + a

TERMINATE:     WRITE n
```

3.13 Use Theorem 3.2 to prove that if $gcd(b, 561) = 1$ then

$$b^{560} \equiv 1 \pmod 3,$$
$$b^{560} \equiv 1 \pmod{11},$$
$$b^{560} \equiv 1 \pmod{17}.$$

It follows from these three congruences that

$$b^{560} \equiv 1 \pmod{561}.$$

3.14 Find the smallest pseudoprime (base 2) which is larger than a thousand. Prove that this number is in fact a Carmichael number.

3.15 Find the value of $\phi(n)$ for $n = 9, 10, 11,$ and 12.

3.16 Write a program to compute $\phi(n)$ by counting one for each $i \leq n$ such that $gcd(i, n) = 1$. List the values of $\phi(n)$ for n up to 200.

3.17 What patterns can you pick out of the table of values of $\phi(n)$?

3.18 Is $\phi(n)$ always even when n is larger than 2? Prove your assertion.

3.19 Prove that for any integer $x > 1$, we have that

$$gcd(x^a - 1, x^b - 1) = x^{gcd(a,b)} - 1.$$

3.20 Let p be a prime and b be an integer. Prove that if q is a prime divisor of $b^p - 1$ then

$$q \quad \text{divides} \quad b - 1 \quad \text{or} \quad q \equiv 1 \pmod{p}.$$

3.21 In 1909, A. Wieferich proved that if p is an odd prime and if there exist non-zero integers x, y, and z not divisible by p and satisfying the Fermat equation

$$x^p + y^p = z^p,$$

then the following congruence holds:

$$2^{p-1} \equiv 1 \pmod{p^2}.$$

Using a computer search, find the smallest prime satisfying this congruence.

4

The RSA Public Key Crypto-System

4.1 The Basic Idea

One of the principal motivations for the flurry of work that has been done on factorization and primality testing over the past decade has been the invention by Rivest, Shamir, and Adleman in 1977 of a "public key crypto-system" based on the fact that multiplication of two large primes is much easier then factoring the resulting product.

The basic idea of a public key or asymmetric encryption scheme was independently proposed by Diffie and Hellman at Stanford and by Merkle at the University of California in 1976. In the codes in use until then, the encoding and decoding keys were effectively identical. This meant that no matter what the computational complexity of the coding scheme might be, it was very vulnerable because there were at least two parties with copies of the key. And in practice one of these would be an operative out in the field where it would be much harder to keep the key secure.

In a public key or asymmetric encryption scheme, the encryption and decryption keys are distinct. Of course, it is possible to figure out the decryption key from the encryption key, at least in theory. The idea is to make this computationally infeasible. If you can measure the time to obtain the decryption key from the encryption key in years, even under an assault by a battery of the world's fastest computers, then your code is effectively secure.

This is called a public key encryption scheme because not only does it eliminate the danger of the encryption scheme being stolen, the encryption scheme can now be freely published so that anyone can send in a coded message. This has obvious business applications.

One can also turn this idea around to obtain a "signature code". Here you publish the decryption key but keep the encryption key secret. If you send a coded message to your bank, for example, and using your decryption key they discover a message to transfer $250 000 to your brother in Toledo, the fact that the decryption key yielded a sensible message means that the message was encoded by someone with the encryption key, which they assume means you.

Several people have suggested different asymmetric or public key encryption schemes. One of the few that has withstood the test of keeping the encryption and decryption schemes computationally quick and yet the passage from one to the other impossible in practice is the scheme by Rivest, Shamir, and Adleman (RSA) developed at M.I.T. in 1977. It is based on Euler's Theorem 3.4 of the last chapter.

Let p and q be distinct large primes and let n be their product. Assume that we also have two integers, d (for decryption) and e (for encryption) such that

$$d \times e \equiv 1 \pmod{\phi(n)}. \tag{4.1}$$

The integers n and e are made public, while p, q, and d are kept secret.

Let M be the message to be sent where M is a positive integer less than and relatively prime to n. If we keep M less than both p and q, then we will be safe. In practice, if is enough to keep M less than n for the probability that a random M is divisible by p or q is so small as to be negligible. A plaintext message is easily converted to a number by using, say,

$$\text{blank} = 99, \quad A = 10, \quad B = 11, \quad \ldots, \quad Z = 35,$$

so that HELLO becomes 1714212124. If necessary, the message can be broken into blocks of smaller messages: 17142 12124.

The encoder computes and sends the number

$$E = M^e \text{ MOD } n \tag{4.2}$$

which we know from Algorithm 3.3 can be done very quickly. To decode, we simply compute

$$E^d \text{ MOD } n.$$

By Theorem 3.4 and our equation (4.1) we have that

$$\begin{aligned} E^d &\equiv (M^e)^d \equiv M^{e \times d} \equiv M^{(\text{multiple of } \phi(n))+1} \pmod{n} \\ &\equiv 1 \times M \equiv M \pmod{n}. \end{aligned} \tag{4.3}$$

Since M and E^d MOD n both lie between 0 and n, they must be equal.

If e has been chosen relatively prime to $\phi(n)$, then the next lemma guarantees that there is a d. The proof of the lemma shows how to find d.

Lemma 4.1 *Given relatively prime integers a and m, there exists an integer b, unique modulo m, such that*

$$a \times b \equiv 1 \pmod{m}.$$

Definition: If $a \times b \equiv 1 \pmod{m}$, then we say that b *is the inverse of a modulo m.*

Proof. By the Euclidean algorithm (Algorithm 1.8) we can find integers b and c such that

$$a \times b + m \times c = 1.$$

This means that $a \times b$ is congruent to 1 modulo m.

Let e be any other integer satisfying

$$
\begin{aligned}
a \times e &\equiv 1 \pmod{m}, \text{ then} \\
e &\equiv e \times (a \times b) \pmod{m} \\
&\equiv (a \times e) \times b \pmod{m} \\
&\equiv b \pmod{m}.
\end{aligned}
$$

Q.E.D.

As we shall prove later in this chapter, if we know the factorization of n, namely $n = p \times q$ where p and q are distinct primes, then we can easily compute $\phi(n)$ by

$$\phi(n) = (p - 1) \times (q - 1). \tag{4.4}$$

There is no simpler way of computing $\phi(n)$. In fact, knowing $\phi(n)$ is equivalent to knowing the factorization because we can find $p + q$:

$$p + q = n - \phi(n) + 1 = p \times q - (p \times q - p - q + 1) + 1, \tag{4.5}$$

and then $p - q$ is

$$
\begin{aligned}
p - q = \sqrt{(p + q)^2 - 4n} &= \sqrt{p^2 + 2p \times q + q^2 - 4p \times q} \tag{4.6} \\
&= \sqrt{p^2 - 2p \times q + q^2},
\end{aligned}
$$

and finally:

$$p = [(p + q) + (p - q)]/2, \quad q = [(p + q) - (p - q)]/2. \tag{4.7}$$

The problem of finding d, the decryption key, has been reduced to finding the factorization of n.

In practice, one takes p and q to be roughly 100-digit primes. Thus their product is about two hundred digits, well beyone current publicly known factorization techniques for "generic" integers. You do have to be somewhat careful in the choice of p and q. As we will see in Chapter 5, one of the very fast factorization techniques looks for prime factors p with the property that all primes dividing $p - 1$ are small, say less than a million. You therefore want to make sure that you choose p and q so that $p - 1$ and $q - 1$ are each divisible by a big prime, say p' and q', respectively.

For reasons that I will not go into, we also want $\phi(\phi(p \times q))$ to be large and divisible by large primes which means that $gcd(p - 1, q - 1)$ should be small and $p' - 1$ and $q'.-1$ should each be divisible by a large prime. Further discussion of how the RSA public key encryption scheme works in practice can be found in the articles referenced at the end of this chapter.

4.2   An Example

To construct your code, start by finding large primes to divide $p' - 1$ and $q' - 1$. We will start with two primes over a million:

$$p'' \;=\; 4\,813\,309, \text{and}$$
$$q'' \;=\; 1\,162\,957.$$

Run through the integers that are one more than an even multiple of p'' and q'', respectively, until you find one that passes the pseudoprime test. Verify that it really is a prime. (For numbers this small, you can use trial division to verify primality.)

$$p' \;=\; 22 \times (4\,813\,309) + 1 \;=\; 105\,892\,799, \quad \text{and}$$
$$q' \;=\; 6 \times (1\,162\,957) + 1 \;=\; 6\,977\,743.$$

Now run through the integers one more than an even multiple of p' and q', respectively, until you find one that passes the pseudoprime test. Verify that it is a prime.

$$
\begin{aligned}
p &= 20 \times (105\,892\,799) + 1 &=& \qquad 21178\,55981, \quad \text{and} \\
q &= 4 \times (6\,977\,743) + 1 &=& \qquad 27\,910\,973, \\
n &= &p \times q =& 59\,11142\,11035\,79513, \\
\phi(n) &= (p - 1) \times (q - 1) &=& 59\,11141\,89578\,12560.
\end{aligned}
$$

Choose an e relatively prime to $p - 1$ and to $q - 1$:

$$e = 123$$

will do nicely. Using Algorithm 1.8 in the manner explained above, find d such that $e \times d \equiv 1 \pmod{\phi(n)}$.

$$d = 18\,26206\,43934\,70547.$$

Note that d must be positive, so that if Algorithm 1.8 returns a negative value then add a multiple of $\phi(n)$ to get a positive value.

The code is now set up. Publish the values of n and e, lock the value of d away in a secure place. The actual primes p and q as well as $\phi(n)$ are no longer needed and it is safest to destroy all trace of them. If we keep our blocks to at most sixteen digits, then each piece of message will be less than n. The odds of an arbitrary integer less than 10^{16} being divisible by p or q is only about 1 in 300 million. As p and q get larger, the odds decrease even further.

The following algorithms are set up for the particular code I have described.

Algorithm 4.2 *This is an encoding program for the* RSA *public-key cryptosystem.*

```
INITIALIZE:   READ message
              number ← CONVERT_TO_NUMBER (message)
              modulus ← 59111421103579513
              exponent ← 123
```

> **message** *is the character string to be encoded.*

```
BREAK_NUMBER:
```

> *Break* **number** *into blocks of at most 16 digits each. Let* b *be the number of blocks and let* block$_i$ *be the* i^{th} *block.*

```
CODE:         FOR i = 1 to b DO
                  code_i ← MODEXPO(block_i,exponent,modulus).
```

```
TERMINATE:    For i = 1 to b DO
                  WRITE code_i
```

```
CONVERT_TO_NUMBER(message):
```

> *Reading* **message** *one character at a time, translate each character to a two-digit number and then concatenate the resulting numbers. Return the concatenated number.*

```
MODEXPO(a,b,m):
```

Use Algorithm 3.3 to compute a^b MOD m. *Return value of* a^b MOD m *to caller.*

Algorithm 4.3 *This is a decoding algorithm for the* RSA *public-key cryptosystem.*

```
INITIALIZE:    READ b, code₁,...,code_b
               modulus  ←  591111421103579513
               exponent  ←  18262064393470547
```

Input the code numbers.

```
DECODE:        FOR i = 1 to b DO
                   block_i  ←  MODEXPO(code_i,exponent,modulus)

FIND_MESSAGE:  number  ←  CONCATENATE(block_i)
               message  ←  CONVERT_TO_CHARACTER(number).
```

Fill each block out to sixteen digits by inserting 0's as needed at the beginning of the block and then concatenate the resulting blocks into a single number. *Then call the* CONVERT_TO_CHARACTER *subroutine.*

```
TERMINATE:     WRITE message.

CONVERT_TO_CHARACTER(number):
```

Reading the number from right to left, convert each pair of digits to a character. Return the resulting string of characters.

```
MODEXPO(a,b,m):
```

Use Algorithm 3.3 to compute a^b MOD m. *Return value of* a^b MOD m *to caller.*

The signature code uses the same two algorithms, but with the values of d and e reversed so that anyone can decode but only the holder of d can encode a message. If a coded message has not been produced by Algorithm 4.2, it is unlikely that running it through Algorithm 4.3 will produce anything that makes sense.

4.3   The Chinese Remainder Theorem

We will conclude this chapter by investigating Euler's ϕ function a little more closely. We need to prove that

$$\phi(p \times q) = (p - 1) \times (q - 1),$$

but while we are at it we will prove a good deal more that will come in handy later. The starting point is an algorithm that appeared in the first century A.D. simultaneously in China, in the writings of Sun-Tsu, and in Judea, in a book by Nichomachus of Gerasa.

Theorem 4.4 (Chinese Remainder Theorem) *Let $m_1, m_2, \ldots, m_r$ be positive integers that are pairwise relatively prime (i.e., no two share a common factor larger than one). Let $a_1, a_2, \ldots, a_r$ be arbitrary integers. Then there exists an integer a such that*

$$
\begin{aligned}
a &\equiv a_1 \pmod{m_1} \\
&\equiv a_2 \pmod{m_2} \\
&\quad \cdot \\
&\quad \cdot \\
&\quad \cdot \\
&\equiv a_r \pmod{m_r}.
\end{aligned}
$$

Furthermore, a is unique modulo $M = m_1 \times m_2 \times \cdots \times m_r$.

Proof: We will actually find an algorithm for constructing a. For each i from 1 up to r, define M_i by

$$M_i = (M/m_i)^{\phi(m_i)}.$$

Since M/m_i is relatively prime to m_i and divisible by m_j for every j not equal to i, we have that

$$
\begin{aligned}
M_i &\equiv 1 \pmod{m_i}, \\
M_j &\equiv 0 \pmod{m_i} \quad \text{for every } j \text{ not equal to } i.
\end{aligned}
$$

Define a by

$$a = a_1 \times M_1 + a_2 \times M_2 + \cdots + a_r \times M_r.$$

To see that a is unique modulo M, let b be any other integer satisfying the r congruences. Then for each m_i, a and b are congruent modulo m_i. In other words, m_i divides $b - a$. Since this is true for every i, M divides $b - a$ which means that

$$a \equiv b \pmod{M}.$$

Q.E.D.

Of course, one need only compute the values of M_i modulo M, which keeps the computations a little more reasonable. An even more efficient algorithm for finding a was discovered by H. L. Garner in 1958.

Algorithm 4.5 *Given a set of pairwise relatively prime moduli:*

$$m_1, m_2, \ldots, m_r$$

and a set of residue classes:

$$a_1, a_2, \ldots, a_r,$$

this computes an integer a satisfying

$$a \equiv a_i \pmod{m_i}, \quad \text{for all} \quad i.$$

INITIALIZE: READ r
                 FOR i = 1 to r DO
                     READ a_i,m_i

FIND_INVERSES: FOR j = 2 to r DO
                     FOR i = 1 to j - 1 DO
                         $c_{i,j}$ ← INVERSE(m_i,m_j)

INVERSE(s, t) *is the inverse of s modulo t.*

FIND_W'S: FOR j = 1 to r DO
                     w_j ← a_j MOD m_j
                     FOR i = 1 to j - 1 DO
                         w_j ← ((w_j - w_i) × $c_{i,j}$) MOD m_j

This subroutine constructs w_j's with the property that

$$w_1 + m_1 \times w_2 + m_1 \times m_2 \times w_3 + \cdots + m_1 \times m_2 \times \cdots \times m_{j-1} \times w_j$$

is congruent to $a_j (mod\ m_j)$.

```
COMPUTE_A:        a ← w_r
                  FOR i = r - 1 to 1 BY -1 DO
                       a ← a × m_i + w_i
```

$$a = w_1 + m_1 \times w_2 + m_1 \times m_2 \times w_3 + \cdots + m_1 \times m_2 \times \cdots \times m_{r-1} \times w_r.$$

```
TERMINATE:        WRITE a

INVERSE(s,t):     u_1 ← 1;  u_2 ← s
                  v_1 ← 0;  v_2 ← t
                  WHILE v_2 ≠ 0 DO
                       q ← ⌊u_2/v_2⌋
                       FOR n = 1 to 2 DO
                            temp = u_n - q × v_n
                            u_n = v_n
                            v_n = temp
                  RETURN u_1
```

This subroutine is Algorithm 1.8 with the second set of variables surpressed. The final value of u *is the inverse.*

4.4 What if the Moduli are not Relatively Prime?

What if two or more of the moduli are divisible by the same prime? Neither form of the Chinese Remainder Theorem works in this case.

We first observe that if we have a congruence with a modulus which is divisible by more than one prime, then we can split our single congruence into several congruences as long as the new moduli are relatively prime and their product is the original modulus. As an example,

$$x \equiv 3 \pmod{45}$$

is equivalent to

$$x \equiv 3 \pmod 5 \quad \text{AND} \quad x \equiv 3 \pmod 9.$$

The equivalence is a consequence of the Chinese Remainder Theorem.

Let us assume that in our system of congruences we have two moduli, say m_1 and m_2, that are both divisible by the same prime, p. We split each of our congruences into two congruences as explained above where one of the new moduli is the highest power of p dividing m_1 or m_2, respectively. As an example, the two congruences:

$$x \equiv 3 \pmod{45},$$
$$x \equiv 7 \pmod{756},$$

are split into four congruences:

$$x \equiv 3 \pmod 9, \qquad x \equiv 3 \pmod 5,$$
$$x \equiv 7 \pmod{27}, \qquad x \equiv 7 \pmod{28}.$$

We now have two congruences that both involve powers of p. One of two things has to happen:

(1) The congruences are contradictory and so there are no solutions. This is the case with the example given above. If $x \equiv 3 \pmod 9$ then $x \equiv 3$, 12, or $21 \pmod{27}$.

or

(2) Both of the congruences for powers of p are implied by the congruence with the higher power. This means we can get rid of one of our equations, leaving us with three congruences with relatively prime moduli for which we can use the Chinese Remainder Theorem.

The next example shows the second possibility:

$$x \equiv 7 \pmod{200},$$
$$x \equiv 82 \pmod{375}.$$

This splits into four congruences:

$$x \equiv 7 \pmod{25}, \qquad x \equiv 7 \pmod 8,$$
$$x \equiv 82 \pmod{125}, \qquad x \equiv 82 \pmod 3.$$

The congruence modulo 25 is a special case of the congruence modulo 125, so we really have three congruences to relatively prime moduli:

$$x \equiv 82 \pmod{125},$$
$$x \equiv 7 \pmod 8, \text{ and}$$
$$x \equiv 82 \equiv 1 \pmod 3,$$

which has as solution

$$x \equiv 1207 \pmod{3000}.$$

4.5 Properties of Euler's ϕ Function

Lemma 4.6 *If $gcd(m, n) = 1$, then*

$$\phi(m \times n) = \phi(m) \times \phi(n).$$

Proof: Let a be a positive integer less than and relatively prime to $m \times n$. In other words, a is one of the integers counted by $\phi(m \times n)$. We consider the correspondence

$$a \rightarrow (a \ \text{MOD} \ m, a \ \text{MOD} \ n).$$

The integer a is relatively prime to m and relatively prime to n, so a MOD m and a MOD n are relatively prime to m and n, respectively. This means that each integer counted by $\phi(m \times n)$ corresponds to a pair of integers, the first counted by $\phi(m)$ and the second counted by $\phi(n)$. By the second part of Theorem 4.4, distinct integers counted by $\phi(m \times n)$ correspond to distinct pairs. Therefore $\phi(m \times n)$ is at most the number of such pairs:

$$\phi(m \times n) \leq \phi(m) \times \phi(n).$$

In the other direction, we take a pair of integers, one counted by $\phi(m)$ and the other counted by $\phi(n)$. Since m and n are relatively prime, we can use the first part of Theorem 4.4 to construct a unique positive integer a less than and relatively prime to $m \times n$.

$$(b, c) \xrightarrow{\text{ChineseRemainderTheorem}} a \equiv b \ (\text{mod} \ m), \equiv c \ (\text{mod} \ n),$$

So the number of such pairs is at most $\phi(m \times n)$:

$$\phi(m \times n) \geq \phi(m) \times \phi(n).$$

Q.E.D.

In Exercise 4.17 you are asked to prove that if p is a prime, then

$$\phi(p^a) = p^{a-1} \times (p - 1).$$

With this equality and Lemma 4.6, we can rapidly calculate $\phi(n)$ for any n which we can factor.

Theorem 4.7 *Let n have the prime factorization*

$$
\begin{aligned}
n &= p_1^{a_1} \times p_2^{a_2} \times \cdots \times p_r^{a_r}, \ then \\
\phi(n) &= p_1^{a_1-1} \times (p_1 - 1) \times p_2^{a_2-1} \times (p_2 - 1) \times \cdots \times p_r^{a_r-1} \times (p_r - 1) \\
&= n \times (1 - \frac{1}{p_1}) \times (1 - \frac{1}{p_2}) \times \cdots \times (1 - \frac{1}{p_r}).
\end{aligned}
$$

REFERENCES

W. Diffie and M. E. Hellman, "New Directions in Cryptography," *IEEE Trans. Inform. Theory*, **IT-22**(1976), 644-654.

P. J. Hoogendoorn, "On a Secure Public-key Cryptosystem," in *Computational Methods in Number Theory*, edited by H. W. Lenstra, Jr. and R. Tijdeman, Mathematisch Centrum, Amsterdam, 1982.

R. C. Merkle, "Secure Communications over Insecure Channels," *Comm. ACM*, **21**(1978), 294-299.

R. L. Rivest, A. Shamir, & L. Adleman, "A Method for Obtaining Digital Signatures and Public-Key Cryptosystems," *Comm. ACM*, **21**(1978), 120-126.

G. J. Simmons, "Cryptology: The Mathematics of Secure Communication," *The Mathematical Intelligencer*, **1**(1979), 233-246.

4.6   EXERCISES

4.1 Prove that if a and m are not relatively prime, then there is no b for which

$$
a \times b \equiv 1 \pmod{m}.
$$

4.2 Given that $n = 19\,74936\,15358\,94833$ and $\phi(n) = 19\,74936\,12325\,17120$, and knowing that n is a product of two primes, find those primes.

4.3 For each of the pairs a, m, find an inverse of a modulo m or show that no such inverse exists:

$$a = \quad 25, \qquad m = \quad 928\ 102$$
$$315, \qquad\qquad 864\ 247$$
$$1001, \qquad\qquad 2\ 671\ 835$$
$$2643, \qquad\qquad 23\ 175\ 586$$
$$5231, \qquad\qquad 33557\ 79009\ .$$

4.4 Decode the following message sent using the code set up in Algorithm 4.2. It is a quotation from Shakespeare's "Hamlet". The coded message consists of four integers

$$39\ 25736\ 57380\ 83976$$
$$8\ 66571\ 70599\ 56870$$
$$14569\ 39934\ 49451$$
$$14\ 57541\ 36754\ 04137$$

4.5 Listed below are the published values of n and e for four public key codes, followed by four signatures. Match each signature to the appropriate code and decipher the signatures. Each signature is given as two integers.

$$n \ = \ \ 17\ 97900\ 80412\ 22471, \qquad e = 101;$$
$$n \ = \ \ 14\ 38977\ 95738\ 78299, \qquad e = 101;$$
$$n \ = \ \ 34\ 59502\ 11241\ 56601, \qquad e = 101;$$
$$n \ = \ \ 15\ 14834\ 44886\ 02009, \qquad e = 101.$$

first signature: 19565 10306 21381
                          30889 96647 52558

second signature: 10 46392 78183 44372
                      11 79772 25496 34348

third signature: 16 19738 57937 34878
                      12 25442 32940 26625

fourth signature: 28 73245 74914 14758
                      5 79586 11521 68412.

4.6 Two people have their own public-key system (m_1, e_1, d_1) and (m_2, e_2, d_2). They have each published the values of m_i and e_i and want to communicate with each other so that the communication will be secure and the recipient will know that the message could only have come from the other person. How can this be done?

4.7 It is much easier to read a large integer when it is broken into blocks of five digits. Write a subroutine that can be included in your programs so that before outputting a number of ten or more digits, the program breaks it into substrings of length at most five, as the numbers in Exercise 4.4 are presented.

4.8 It is much easier to see that a large integer has been entered correctly if it can be entered in blocks with blanks between them. Write a subroutine that will check if an inputted number includes blanks and if so will concatenate (push everything together).

4.9 Find the decoding key d for the code whose published values of n and e are

$$n = 233\,570\,063, \quad e = 125.$$

4.10 Construct your own code such that $p' - 1$ and $q' - 1$ each have a prime factor over a million and p and q are 12 or 13-digit primes. Only hand in your values for n and e. Keep the value of d secret. As you learn more about factorization, try to break each other's codes.

4.11 Using Algorithm 4.5 find the smallest positive integer which satisfies the following system of congruences:

$$
\begin{aligned}
a &\equiv 2 \pmod{21} \\
&\equiv 3 \pmod{31} \\
&\equiv 6 \pmod{61} \\
&\equiv 10 \pmod{101} \\
&\equiv 15 \pmod{151} \\
&\equiv 31 \pmod{311} \\
&\equiv 43 \pmod{431}.
\end{aligned}
$$

4.12 Find the smallest positive integer which satisfies each of the following systems of congruences or prove there is no solution:

$$
\begin{aligned}
x &\equiv 22 \pmod{441} \\
&\equiv 36 \pmod{455};
\end{aligned}
$$

$$
\begin{aligned}
y &\equiv 16 \pmod{303} \\
&\equiv 25 \pmod{378} \\
&\equiv 13 \pmod{423};
\end{aligned}
$$

$$
\begin{aligned}
z &\equiv 25 \pmod{275} \\
&\equiv 13 \pmod{495} \\
&\equiv 46 \pmod{616}.
\end{aligned}
$$

4.13 Modify Algorithm 4.5 so that it will automatically handle systems of congruences where the moduli are not relatively prime.

4.14 Four 12-hour clocks are lined up. The first reads 10:41 and loses 2 hours and 24 minutes a day. The second reads 4:50 and loses 45 minutes a day. The third reads 3:45 and gains 1 hour and 20 minutes a day. The fourth reads 1:05 and keeps perfect time. How long will it be before all four clocks read exactly the same time?

4.15 In Algorithm 4.5, explain why the recursive definition of w_j gives us the congruence:

$$w_1 + m_1 \times w_2 + \cdots + m_1 \times \cdots \times m_{j-1} \times w_j$$

$$\equiv a_j \pmod{m_j}.$$

4.16 In Algorithm 4.5, explain why

$$w_1 + m_1 \times w_2 + \cdots + m_1 \times \cdots \times m_{r-1} \times w_r$$

is congruent to $a_j \pmod{m_j}$ for every j.

4.17 Let p be a prime. How many of the positive integers less than or equal to p^a are divisible by p? Use this to prove that

$$\phi(p^a) = p^{a-1} \times (p - 1).$$

4.18 Find two positive integers m and n which are not relatively prime but for which

$$\phi(m \times n) = \phi(m) \times \phi(n),$$

or prove it cannot be done.

4.19 Write a program using trial division and Theorem 4.7 to compute $\phi(n)$. Use it to compute $\phi(n)$ for each of the following values of n:

$$51\ 005$$
$$107\ 653$$
$$1\ 294\ 704$$
$$1\ 494\ 108$$
$$614\ 739\ 125$$

4.20 Use the results from Exercise 3.16 to find all integers n for which $\phi(n) = 12$. Prove that there are no integers n larger than 200 for which $\phi(n)$ is 12.

5

Factorization Techniques from Fermat to Today

> "The term *Science* should not be given to anything but
> the aggregate of the recipes that are always successful.
> All the rest is *literature*."
> – Paul Valéry

5.1 Fermat's Algorithm

Our only factorization algorithm so far is Algorithm 2.4, which will work
fine for numbers up to ten or eleven digits, but quickly bogs down after
that. Part of the problem with trial division is that it does too much. It
is not only a factorization algorithm, it will also prove primality as long
as you have the time to test for divisibility up to the square root of the
number in question.

But we are now equipped with the strong pseudoprime test which means
that before we even start looking for factors we know that the number in
question is composite. Also, we are willing to settle for an algorithm that
does not spit out all the prime divisors but rather just breaks our integer
into a product of two integers

$$n = a \times b,$$

where a and b are each larger than 1. We can now test a and b for primality.
If either is composite, we break it into two factors and fairly quickly we
will have n reduced to a product of primes.

The first of the modern algorithms we will look at is due to Fermat. It is
not usually implemented these days unless it is known that the number to
be factored has two factors which are relatively close to the square root of
the number. But it does contain the key idea behind two of today's most
powerful algorithms for factoring numbers with large prime factors, the
Quadratic Sieve and the Continued Fraction Algorithms.

Fermat's idea is the following. If n is the number to be factored and if n
can be written as a difference of two perfect squares:

$$n = x^2 - y^2,$$

then

$$n = (x - y) \times (x + y),$$

and we have succeeded in breaking n into two smaller factors. Furthermore, if we assume that the n we start with is odd (a safe assumption), then every representation of n as a product of two integers arises in this way.

To see this, let $n = a \times b$, where a and b are odd because n is odd. Let

$$x = (a + b)/2 \quad \text{and} \quad y = (a - b)/2.$$

Then

$$x^2 - y^2 = (a^2 + 2 \times a \times b + b^2 - a^2 + 2 \times a \times b - b^2)/4 = a \times b = n.$$

Fermat's algorithm works in the opposite direction from trial division. In Algorithm 2.4 we started by looking for small factors and worked our way up to the square root of n. Here we start by looking for factors near the square root of n and work our way down.

Given a positive odd integer n to be factored, we search for integers x and y such that $x^2 - y^2 = n$. We start with x equal to the smallest integer greater than or equal to the square root of n and try increasing y's until $x^2 - y^2$ either equals or is less than n. In the first case we are done, in the second we increase x by one and iterate. We continue until we have success. If we set r equal to $x^2 - y^2 - n$, then we have success when $r = 0$.

This algorithm is further stream-lined by keeping track of $u = 2x+1$ and $v = 2y+1$ instead of x and y. All we are really interested in is keeping track of r. The variable u tracks the amount r increases when x^2 is replaced by $(x + 1)^2$, v tracks the amount r decreases when y^2 is replaced by $(y + 1)^2$. As x and y increase by one, u and v increase by 2.

Algorithm 5.1 *Fermat's algorithm to find a factor of n near its square root.*

```
INITIALIZE:   READ n
              sqrt ← ⌈√n⌉
              u ← 2 × sqrt + 1
              v ← 1
              r ← sqrt × sqrt - n
```

sqrt is the smallest integer greater than or equal to $\sqrt{n}$. The initial value of x is sqrt. *The initial value of y is 0. See Exercise 5.7 for a subroutine that will compute the greatest integer less than or equal to $\sqrt{n}$.*

```
X_LOOP:        WHILE r ≠ 0 DO
                    IF r > 0 THEN CALL Y_LOOP
                    IF r < 0 THEN DO
                         r ← r + u
                         u ← u + 2
```

If r is negative then we increase x by 1.

```
TERMINATE:     a ← (u + v - 2)/2
               b ← (u - v)/2
               WRITE a, b
```

x − y and x + y are now two factors of n.

```
Y_LOOP:        WHILE r > 0 DO
                    r ← r - v
                    v ← v + 2
               RETURN
```

If r is positive then we increase y by 1. This decreases r by v. We then reset the value of v. When r is no longer positive, we return the current values of r and v.

This algorithm has some nice features, chief among them that the loops involve no multiplication or division so that they cycle extremely quickly. The problem, of course, is the prodigious number of cycles required. To factor

$$17836\ 47329 = 84\ 449 \times 21\ 121$$

requires 10551 cycles of X_LOOP and 31664 cycles of Y_LOOP.

There are some techniques for speeding up this algorithm. In particular, if $x^2 - n$ is not a perfect square when we start Y_LOOP, then r will not be 0 when we leave it (see Exercise 5.6). In the next two chapters, we will be developing tests that can quickly tell us if an integer is probably a perfect square or definitely not a perfect square, and these can be incorporated.

But this algorithm still suffers from the same defect as trial division, it will prove primality if you let it run long enough. If run on a prime number n, it will eventually come up with the factors 1 and n, and since there are no factors of n closer to its square root, n must be prime. Incidentally, this is a terrible way to prove primality as the total number of cycles required in n minus the square root of n, much worse than proving primality by trial division.

5.2   Kraitchik's Improvement

Maurice Kraitchik (1882-1957) realized that a major saving of time could be accomplished if instead of looking for x and y satisfying $x^2 - y^2 = n$, we settle for "random" x and y satisfying

$$x^2 \equiv y^2 \pmod{n}.$$

Finding such a pair (x, y) no longer guarantees us a factorization. But it does mean that n divides

$$x^2 - y^2 = (x - y) \times (x + y),$$

and you now have at least a 50-50 chance that the prime divisors are distributed among the divisors of both of these factors so that the *gcd* of n and $x - y$ will be a nontrivial factor of n. That is to say, the *gcd* will be neither 1 nor n.

His approach to finding such pairs (x, y) was rather *ad hoc*. A few years later, in 1931, D. H. Lehmer and R. E. Powers showed how to find these pairs systematically by using continued fractions. Their algorithm, however, was not particularly practical until the coming of high speed computers. By the late 1960s and early 1970s the computer hardware had advanced to the point where people realized that the Lehmer-Powers algorithm was worth re-examination. Daniel Shanks was one of the first to come up with a practical algorithm using continued fractions and Kraitchik's idea, the Square Forms Factorization (SQUFOF). In 1975, John Brillhart and Michael Morrison, using continued fractions more along the lines of Lehmer and Powers than Shanks, published what has become the standard form of the continued fraction algorithm (CFRAC) and started the current cottage industry of factoring truly big numbers.

For about a decade, the Brillhart-Morrison CFRAC reigned as the fastest means of factoring large numbers with large prime factors. It is still in use today. The "Georgia Cracker", built at the University of Georgia only a few years ago for the sole purpose of factoring numbers, runs on this algorithm and is handling roughly 60-digit integers. But the past few years have also seen CFRAC supplanted by Carl Pomerance's Quadratic Sieve (QS) and Peter Montgomery's refinement, the Multiple Polynomial Quadratic Sieve (MPQS). These employ a different approach to the problem of finding pairs (x, y) for which $x^2 \equiv y^2 \pmod{n}$, an approach using large amounts of memory. Their ascendance is directly the result of the advent of large, cheap memory. They are also particularly amenable to parallel processing.

5.3   Pollard Rho

One drawback to both the Continued Fraction and Quadratic Sieve methods is that they are not any faster on finding moderately sized factors, say around 10^5 to 10^{10}, than they are on finding the really big factors. If a composite number has a moderately sized prime divisor, for example,

$$1\,888\,129,$$

which is too large to find by trial division, there should still be a quick way of finding it.

I will be describing not one but two such algorithms, both due to J. M. Pollard, the first published in 1975 and the second in 1974. While they will not usually work if all the prime factors are big (larger than 10^{12}), they are very simple to understand and easy to program. In practice, after exhausting trial divisors up to 10^4 or 10^5, one runs the Pollard tests for a while before pulling out the really big guns of the Elliptic Curve Method (Chapter 14), CFRAC (Chapter 11), or a Quadratic Sieve Algorithm (Chapter 8).

The first Pollard algorithm was named the Monte Carlo Method by him because of its pseudo-random nature. For reasons that will become clear as I explain it, it is now more popularly known as Pollard rho. Let n be a composite number and d an unknown nontrivial divisor of n. Let $f(x)$ be a simple irreducible (cannot be factored) polynomial in x. In practice we will use $x^2 + 1$ or something similar. Starting with an integer x_0, we create a sequence from the recursive definition:

$$x_i = f(x_{i-1}) \text{ MOD } n.$$

If $x_0 = 2$, $f(x) = x^2 + 1$ and $n = 1133$, our sequence will be

$$
\begin{aligned}
x_0 &= & 2 \\
x_1 &= & 5 \\
x_2 &= & 26 \\
x_3 &= & 677 \\
x_4 &= & 598 \\
x_5 &= & 710 \\
x_6 &= & 1049 \\
x_7 &= & 259 \\
x_8 &= & 235 \\
& \vdots &
\end{aligned}
$$

Let

$$y_i = x_i \text{ MOD } d.$$

If we choose $d = 11$ in our example, then the sequence of y_i's is

$$y_0 = 2$$
$$y_1 = 5$$
$$y_2 = 4$$
$$y_3 = 6$$
$$y_4 = 4$$
$$y_5 = 6$$
$$y_6 = 4$$

.

.

.

Since $x_i \equiv f(x_{i-1})$ (mod n), y_i is congruent to $f(y_{i-1})$, modulo d. There are only a finite number of congruence classes, modulo d (namely d of them) and so eventually we will have

$$y_i = y_j,$$

for some pair (i, j). But once that happens, we will keep cycling and for all positive t:

$$y_{i+t} = y_{j+t}. \tag{5.1}$$

Our sequence of y_i's looks like a circle with a tail. In other words, it looks like the greek letter rho, giving rise to the name of this algorithm.

If y_i equals y_j, then

$$x_i \equiv x_j \ (\text{mod } d),$$

and so d divides $x_i - x_j$. There is an excellent chance that x_i and x_j are not equal, and if this is the case then

$$gcd(n, x_i - x_j)$$

is a non-trivial divisor of n.

The problem is that since we do not know d, we do not know the values of the y_i's, and so we do not know when y_i equals y_j. Equation (5.1) comes to our rescue here. There are in fact infinitely many pairs (i, j) for which y_i and y_j are equal. If the length of the cycle is c, then once we are off the tail, any pair (i, j) for which c divides $j - i$ will work. We find some systematic way of choosing a lot of pairs (i, j), and for each pair compute $gcd(n, x_i - x_j)$.

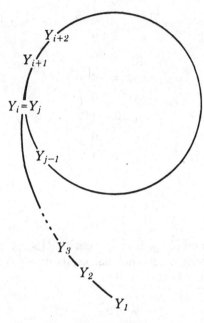

FIGURE 5.1.

The form of the algorithm which we will be using is due to R. P. Brent (1980). In order to avoid storing many values of the x_i's, he suggested looking at the differences:

$$x_1 \quad - \quad x_3$$

$$x_3 \quad - \quad x_6$$
$$x_3 \quad - \quad x_7$$

$$x_7 \quad - \quad x_{12}$$
$$x_7 \quad - \quad x_{13}$$
$$x_7 \quad - \quad x_{14}$$
$$x_7 \quad - \quad x_{15},$$

and in general:

$$x_{2^n-1} - x_j, \quad 2^{n+1} - 2^{n-1} \leq j \leq 2^{n+1} - 1.$$

What is important is the difference between coordinates, which just increases by one each time. We keep moving the smaller coordinate up to guarantee that we get off the tail.

Since we typically have to compute many *gcd*'s, usually thousands or tens of thousands, we get a substantial savings in time if we take the product of, say, ten successive values of $(x_i - x_j)$ MOD n and then take the *gcd* of that product with n. If the *gcd* turns out to be n, we may want to back up over those last ten values and take their *gcd*'s one at a time with n. In practice, however, if n divides the product of ten successive differences, it often divides exactly one of the differences, and your best bet is to start all over with a different polynomial $f(x)$.

WARNING: Never plug a number into either Algorithm 5.2 or 5.3 unless you have run the pseudoprime test and *know* your number is composite. This algorithm can take a very long time if a prime is entered. For safety and convenience, it is a good idea to have this program self-interrupt every so often, say every 1000 or 10000 cycles, and ask if you really want it to continue. Even if the input is composite, there is no guarantee that this algorithm will produce the factorization in your lifetime.

Algorithm 5.2 *Brent's version of the Pollard rho factorization algorithm. The recursion is given by* $f(x_{i+1}) = x_i^2 + c$.

```
INITIALIZE:     READ n, c, max
                x₁ ← 2
                x₂ ← 4 + c
                range ← 1
                product ← 1
                terms ← 0
```

> **n** *is the integer to be factored.* **max** *is the maximum number of cycles before aborting.* **range** *is the number of times we use the current value of* **x₁** *before resetting.* **product** *is the product of up to the last ten differences.*

```
COMPUTE_DIFF:   WHILE terms ≤ max DO
                    FOR j = 1 to range DO
                        x₂ ← (x₂ × x₂ + c) MOD n
                        product ← product × (x₁ - x₂) MOD n
                        terms ← terms + 1
                        IF terms MOD 10 = 0 THEN CALL CHECK_GCD
                    CALL RESET
```

> **x₁** *is held fixed while* **x₂** *increases.*

TERMINATE:

> *If a factor has not been found, you may wish to prompt for a new value of c and then re-initiate the program.*

CHECK_GCD:
```
g ← GCD(n,product)
IF g > 1 THEN DO
        WRITE g
        CALL TERMINATE
product ← 1
RETURN
```

> *If the gcd is larger than 1 then either g is a proper divisor or this algorithm does not work with this value of c. Otherwise return to caller with product reset to 1.*

GCD(a,b):

> *Use Algorithm 1.7 to compute the gcd(a,b). Return this value to caller.*

RESET:
```
x₁ ← x₂
range ← 2 × range
FOR j = 1 to range DO
        x₂ ← (x₂ × x₂ + c) MOD n
RETURN
```

> *Reset x_1 to the last value of x_2, double the range, and then find the next value of x_2. Return these new values to the caller.*

When implementing this algorithm, it is also a good idea not to start out with 2 and 5, but rather to run out the sequence for a while, to give yourself a better chance of getting off the tail and onto the cycle.

In general, you can expect the number of cycles needed to be about the square root of the smallest prime dividing n. That is because the sequence of y_i's usually behave as if they were random. The probability that there will be no repetitions among the first t terms of the sequence is thus

$$\frac{d-1}{d} \times \frac{d-2}{d} \times \cdots \times \frac{d-(t-1)}{d},$$

and this probability drops below 50% right around where t reaches the square root of d.

Two things can go wrong with this algorithm. One is that the first gcd larger than one actually is n. This just means you have been very unlucky in your choice of the polynomial $f(x)$ (or n really is a prime, check your strong pseudoprime test). Pick a different polynomial and run it again.

The other thing that can go wrong is that this algorithm could need a *very* long time to find a divisor. While the cycle length is expected to be around the square root of the smallest prime divisor, it can be as large as the smallest prime divisor. There is also the problem that you usually will not know the size of the smallest prime divisor of n. If in fact it is very large, the y_i's will have a cycle length that is very long. You should be prepared to abort and try a different $f(x)$ or an entirely different algorithm.

5.4   Pollard $p - 1$

This is very similar to the previous Pollard algorithm, although it uses Theorem 3.1. Let us suppose that the number n to be factored has a prime factor p with the property that the primes dividing $p - 1$ are small, say less that 10000. Actually, we will work with the slightly greater restriction that $p - 1$ divides 10000!. Since exponentiation modulo n is so fast, we can compute

$$m = 2^{10000!} \text{ MOD } n$$

fairly quickly:

$$2^{10000!} = (\cdots (((2^1)^2)^3)^4 \cdots)^{10000}.$$

By Theorem 3.1, since $p - 1$ divides 10000!, m is congruent to 1 modulo p, so that p divides $m - 1$. Again, there is an excellent chance that n does not divide $m - 1$, so that

$$g = gcd(m - 1, n)$$

will be a non-trivial divisor of n.

Note that there is nothing special about 2. The same observations hold for $c^{10000!}$ provided that c is relatively prime to n.

In practice, we do not know how close we have to get to 10000 before we have picked up the first prime divisor of n. And we do not want to go so far that we pick them all up. For that reason, we periodically check the value of $gcd(c^{k!} - 1, n)$. If it is still 1, we continue. If it is n, then we have picked up all the divisors of n and we either need to backtrack, try a different value of c, or try a different algorithm. Of course, if it is anything other than 1 or n, then we have found the nontrivial divisor we were looking for.

As in the Pollard rho, we assume that the number n to be factored is known to be composite from a pseudoprime test and does not have any small divisors.

Algorithm 5.3 *Pollard's $p-1$ factorization algorithm.*

```
INITIALIZE:    READ n, c, max
               m ← c


EXPONENTIATE:  FOR i = 1 to max DO
                   m ← MODEXPO(m,i,n)
                   IF i MOD 10 = 0 THEN CALL CHECK_GCD


TERMINATE:
```

If a factor has not been found, you may wish to prompt for a new value of c and then re-initiate the program.

```
MODEXPO(m,i,n):
```

Use Algorithm 3.3 to compute m^i MOD n. Return this value to the caller.

```
CHECK_GCD:     g ← GCD(m-1,n)
               IF g > 1 THEN DO
                       WRITE g
                       CALL TERMINATE
               RETURN
```

If the gcd is larger than 1, then either g is a proper divisor or this algorithm does not work with value of c.

```
GCD(a,b):
```

Use Algorithm 1.7 to compute gcd(a,b) and then return this value to caller.

There is some wasted effort in this algorithm as presented. The exponent 10000! contains much higher powers of the small primes than we are ever likely to need. One frequently suggested improvement is to list the primes which are less than or equal to 10000 and for each such prime p we let e take on the value p a number of times equal to the greatest integer less than or equal to

$$\frac{\log 10000}{\log p},$$

This will cut the number of exponentiations required by about a factor of eight.

This algorithm has the same problems as the previous one. The gcd might equal n, in which case we can go back and change the base 2 to a different integer. It also might crank forever if $p - 1$ has only large prime factors.

If p is the smallest prime dividing n, then the number of cycles required by Algorithm 5.3 is usually the largest prime dividing $p - 1$. Exercises 2.12 and 2.13 should have revealed that the largest prime factor of an arbitrary integer usually falls around the 0.63 power of that integer, so that Algorithm 5.3 with max = 10000 will usually find any prime factors that are less than two million. Of course, there is a fairly wide distribution of the size of the largest prime divisor. Sometimes much larger prime divisors can be found. And there are "small" prime divisors that this algorithm will miss.

Finally, the Pollard $p - 1$ algorithm is one of the reasons for the restrictions on the primes p and q in the RSA public key crypto-system. If $p - 1$ or $q - 1$ have only small prime factors, then Pollard $p - 1$ will crack the code very quickly.

5.5 Some Musings

There is a deep philosophical difference between the approach of Algorithms 2.4 and 5.1 and the approach taken in Algorithms 5.2 and 5.3. The first two factorization algorithms are systematic searches for divisors, what are sometimes called deterministic algorithms. Because of their methodical nature, they are easy to analyze, and one can say exactly how long it will take to find a divisor of a given size.

Algorithms 5.2 and 5.3 are called probabilistic algorithms. We have started to introduce chance and randomness into our procedures. We can now no longer be certain of finding a factor of a given size within a fixed amount of time. But in exchange for that sacrifice we gain that "usually" a factor will be found in much less time than a deterministic algorithm would take.

All of our factorization algorithms from here on will be probabilistic. You simply cannot factor 20 to 30 digit numbers, much less 80 to 100 digit numbers, with deterministic methods. One of the things this means is that we have to be prepared to be unlucky on occasion. If we run a Pollard rho or p-1 test and do not turn up any prime divisors that might be because there are no prime divisors in the appropriate interval or it might be because of bad luck.

The challenge is to learn how to change our luck. All of the probabilistic algorithms have parameters that can be varied. In Algorithm 5.2 it is the function used to generate the sequence: $x^2 + 1$. We can replace this with almost any irreducible quadratic polynomial, say $x^2 + 2$ or $x^2 + 3$ ($x^2 - 2$ is one of the few that will not work). In Algorithm 5.3 it is the base which can be varied. Instead of taking powers of 2 we can take powers of 3 or 5. And, of course, the reason I have shown you both Algorithm 5.2 and 5.3 is that if one of them does not seem to be working, I want you to try the other one. In Chapter 12 you will be seeing still another algorithm, Williams $p + 1$, which looks for prime divisors of the same size as those sought by Pollard's tests.

How long do you keep looking for these middling sized factors before pulling out something like the Quadratic Sieve? That is still more an art than a science, but keep in mind that the algorithms of the next level up are much more cumbersome and it is worth spending at least a few minutes trying to vary your luck first. Theoretically and experimentally, it has been shown that you have a better chance of finding your mid-sized factors if you run several algorithms with several choices of parameters rather than spending the same amount of time grinding out a single algorithm with a single choice of parameter.

The whole problem of analyzing running times and optimum strategies is much more difficult with the probabilistic algorithms than with the deterministic. So much depends on the likely distribution of the number of prime factors and their sizes. Definitive answers here depend on very deep and difficult results on the distribution of primes. Much of what is needed has still not been proven. For the interested reader, I recommend Carl Pomerance's articles listed in the references at the end of this chapter.

Finally, I would like to point out that probabilistic factoring algorithms are inherently unsuitable for saying anything about primality. There is usually no way of distinguishing between consistent bad luck and a prime input. Fortunately we have the pseudoprime tests so that there is no excuse for ever feeding a prime to a probabilistic algorithm. But on a practical level what this means for us is that the problems of factorization and primality testing are now completely divorced. No longer will we see algorithms that can both factor and prove that the factors are prime.

REFERENCES

Richard P. Brent, "An Improved Monte Carlo Factorization Algorithm," Nordisk Tidskrift for Informationsbehandling (BIT), **20**(1980), 176-184.

M. Kraitchik, *Théorie des nombres*, Gauthier-Villars, Paris, 1926.

M. Kraitchik, *Recherches sur la théorie des nombres*, Gauthier-Villars, Paris, 1929.

D. H. Lehmer and R. E. Powers, "On factoring large numbers," *Bull. Amer. Math. Soc.*, **37**(1931), 770-776.

Michael A. Morrison and John Brillhart, "A Method of Factoring and the Factorization of F_7," *Math. Compu.*, **29**(1975), 183-205.

J. M. Pollard, "Theorems on Factorization and Primality Testing," *Proc. Camb. Philo. Soc.*, **76**(1974), 521-528.

J. M. Pollard, "A Monte Carlo Method for Factorization," Nordisk Tidskrift for Informationsbehangling (BIT), **15**(1975), 331-334.

Carl Pomerance, "Analysis and Comparison of some Integer Factoring Algorithms," pp. 89-139 in *Computational Methods in Number Theory, Part I*, H. W. Lenstra, Jr. and R. Tijdeman, eds., Mathematical Centre Tract #154, Mathematisch Centrum, Amsterdam, 1982.

Carl Pomerance, "Lecture Notes on Primality Testing and Factoring," notes by Stephen M. Gagola, Jr., Mathematical Association of America Notes, Number 4, Washington, D.C., 1983.

Hans Riesel, *Prime Numbers and Computer Methods for Factorization*, second printing, Birkhauser, Boston, 1987.

Daniel Shanks, "Class number, a theory of factorization, and genera," pp. 415-440 in *Proceedings of Symposia in Pure Mathematics, Vol. 20*, American Mathematical Society, Providence, R.I., 1971.

5.6   EXERCISES

5.1 Why can we assume that the n we insert into Fermat's Algorithm is odd?

5.2 How many cycles of Algorithm 5.1 will it take to produce the factorization

$$90658\ 64569 = 66\ 457 \times 136\ 417 \quad ?$$

5.3 If p and q are primes, $p < q$, find a formula for how many cycles of Algorithm 5.1 it will take to produce the factorization

$$n = p \times q.$$

5.4 In the X_LOOP of Algorithm 5.1, if r is negative then we replace r by $r + u$. Prove that this new value of r cannot be negative.

5.5 In Algorithm 5.1, why do we not need to reset y back to 0 (reset v to 1) when we leave the Y_LOOP?

5.6 In Algorithm 5.1, why is it that if $x^2 - n$ is not a perfect square when we enter the Y_LOOP then r will not be 0 when we leave that loop?

5.7 Prove that the following algorithm computes $\lfloor \sqrt{n} \rfloor$, the greatest integer less than or equal to the square root of n:

```
INITIALIZE:     READ n
                a ← n
                b ← ⌊(n + 1)/2⌋

MYSTERY_LOOP:   WHILE b < a DO
                    a ← b
                    b ← ⌊(a × a + n)/(2 × a)⌋

TERMINATE:      WRITE a
```

5.8 What happens to Algorithm 5.1 if the value computed as $sqrt$ is in fact less than the square root of n? What happens if it is larger than the smallest integer above the square root of n?

5.9 Use Algorithm 5.1 to factor

$$
\begin{array}{ccc}
19 & 931 & 831 \\
392 & 583 & 509 \\
& 24518 & 39867 \\
& 27863 & 02931 \\
1\ 32883 & 40509 &
\end{array}
$$

5.10 Let n be composite and x and y random integers satisfying

$$x^2 \equiv y^2 \pmod{n}.$$

Explain why there is at least a 50-50 chance that $gcd(n, x - y)$ is a non-trivial factor of n.

5.11 Approximately how many primes are there which are less than two million? If we stored them in memory and used them to run a trial division algorithm, approximately how long would it take to discover the prime factor

$$1\,888\,129?$$

5.12 In the explanation of the Pollard rho algorithm, explain why

$$y_i \equiv f(y_{i-1}) \pmod{d}.$$

5.13 In the explanation of the Pollard rho algorithm, explain why $y_i = y_j$ implies that

$$x_i \equiv x_j \pmod{d}.$$

5.14 Consider the following algorithm for factoring: Given a composite integer n, we choose a random integer r between 1 and n and compute $g = gcd(r, n)$. If g is 1 or n then we choose another random integer and repeat. If it is anything else, then we have found a non-trivial divisor of n. How does this algorthm differ from Algorithm 5.2 and in particular, what are the respective number of cycles you would anticipate if

$$1\,888\,129$$

is in fact the smallest prime dividing n?

5.15 Choose 100 consecutive integers between 500 and 2000 and for each one, call it n, generate the sequence

$$
\begin{aligned}
y_0 &= 2 \\
y_1 &= (2^2 + 1) \text{ MOD } n \\
&\quad . \\
&\quad . \\
&\quad . \\
y_{i+1} &= (y_i^2 + 1) \text{ MOD } n
\end{aligned}
$$

until one of the values repeats. What is the index of the first repeated value? What is the cycle length (that is to say, what is the difference of the indices at which the repeated value occurs)? Describe the distribution of the cycle lengths. How close are the cycle lengths to the square root of your values of n?

5.16 Use Algorithm 5.2 to factor

$$78\ 59947\ 71137$$
$$95\ 01613\ 33249$$
$$250\ 67411\ 91739$$
$$22779\ 31950\ 71137$$
$$26587\ 02640\ 98379$$

5.17 For each of the following values of d, find the smallest integer t for which

$$\frac{d-1}{d} \times \frac{d-2}{d} \times \cdots \times \frac{d-(t-1)}{d} < \frac{1}{2};$$

$d = 25$, $d = 100$, $d = 365$, $d = 10\ 000$, $d = 1\ 000\ 000$.

5.18 Approximately how many digits are there in the integer 10 000!?

5.19 What is the highest power of 2 that divides 10 000!?

5.20 Explain why it is that if $p - 1$ divides 10000! then p divides

$$2^{10000!} - 1.$$

5.21 In theory, which algorithm, 5.2 or 5.3, needs more cycles to find a given prime factor? Explain.

5.22 If i runs to 100 000 in Algorithm 5.3, how large a prime factor can you usually expect to pick up?

5.23 Use Algorithm 5.3 to factor the integers given in Exercise 5.16. Compare running times.

5.24 Find the decoding key for each of the codes given in Exercise 4.5.

6

Strong Pseudoprimes and Quadratic Residues

> "When you have to study a large number, you should begin by finding some of its quadratic residues."
> – Maurice Kraitchik

6.1   The Strong Pseudoprime Test

Having pursued some of the consequences of Fermat's observation, I now invite you to delve back into the mysteries of the structure of the integers. In these next two chapters we will be pulling out a few more jewels that we can apply to our twin problems of factorization and primality testing.

We begin by taking a closer look at the pseudoprime test described in Chapter 3. It has several drawbacks that we are now going to address. One of them is that pseudoprimes, while relatively rare, are not as uncommon as we would like. More troublesome is the presence of the Carmichael numbers which tell us that we can never prove primality by using the pseudoprime test. The following stronger test was developed by Pomerance, Selfridge, and Wagstaff in 1980.

Let us examine an integer, say n, which passes the pseudoprime test for the base b, where b and n have been verified to be relatively prime. This means that n divides

$$b^{n-1} - 1.$$

We can certainly assume that n is odd, otherwise we would not be wasting time trying to decide if it is prime. We can write

$$n = 2m + 1.$$

So n divides

$$b^{2m} - 1 = (b^m - 1) \times (b^m + 1),$$

If n really is a prime, then by Theorem 1.1 it divides at least one of the factors on the right-hand side. And it cannot divide both of them because then it would divide their difference

$$(b^m + 1) - (b^m - 1) = 2.$$

So if n really is a prime, then

$$b^m \equiv 1 \text{ or } -1 \text{ (mod } n). \tag{6.1}$$

On the other hand, if n is composite, there is a fair chance that some of the factors making up n divide $b^m + 1$ while other factors divide $b^m - 1$. In this case, n would pass the pseudoprime test base b, but it would not satisfy Equation (6.1).

We take as an example the first pseudoprime for the base 2:

$$341 = 11 \times 31.$$

In this case, $m = 170$. It is quickly calculated that

$$2^{170} \equiv 1 \text{ (mod } 341).$$

Our number 341 is still looking like a prime. But we are not done yet. That exponent is even and this last equation means that 341 divides

$$2^{170} - 1 = (2^{85} - 1) \times (2^{85} + 1).$$

If 341 were prime, then we would have that

$$2^{85} \equiv 1 \text{ or } -1 \text{ (mod } 341).$$

Now we have it because, in fact

$$2^{85} \equiv 32 \text{ (mod } 341).$$

What has happened is that 11 divides $2^{85} + 1$ while 31 divides $2^{85} - 1$.

In general, let n be a candidate for primality which is relatively prime to b and has passed the pseudoprime test modulo b. Write n as

$$n = 2^a \times t + 1,$$

where t is odd and a is at least 1. Then

$$b^{n-1} - 1 = (b^t - 1) \times (b^t + 1) \times (b^{2t} + 1) \times (b^{4t} + 1) \times \cdots \times (b^{2^{a-1} \times t} + 1), \tag{6.2}$$

and if n really is a prime, then it divides exactly one of these factors.

Definition: An odd integer n is said to be a *strong pseudoprime for the base b* if it is composite, relatively prime to b, and divides one of the factors on the right-hand side of Equation (6.2).

This can be put into a very efficient algorithm that essentially runs as fast as our old pseudoprime test. It is assumed that trial division has been attempted first so that for any small base b which we choose, we know that n, the number to be tested, and b are relatively prime.

Algorithm 6.1 (Strong Pseudoprime Test.)

INITIALIZE: READ n, b

> n *is the integer to be tested.* b *is any positive integer relatively prime to* n.

REDUCE_N-1:   t ← n - 1
                 a ← 0
                 WHILE t is even DO
                     t ← t/2
                     a ← a + 1

> *Find* t *and* a *satisfying:* $n - 1 = 2^a \times t$, t *odd.*

TEST_N_BASE_B: test ← MODEXPO(b,t,n)
                 IF test = 1 or n - 1 THEN CALL PASSED_TEST
                 FOR i = 1 to a - 1 DO
                     test ← (test × test) MOD n
                     IF test = n - 1 THEN CALL PASSED_TEST
                 pass ← 0

> *We check the congruence class of* $b^{2^i t}$ *modulo* n *for each* i *from 0 to* $a - 1$. *If any one is correct, then* n *passes. Otherwise,* n *fails.*

TERMINATE: WRITE pass

> n *passes if and only if* pass *is 1.*

PASSED_TEST: pass ← 1
                 CALL TERMINATE

Strong pseudoprimes do exist. But they are pretty scarce. The first strong pseudoprime for the base 2 is 2047, and as the following table shows, strong pseudoprimes for the base 2 are much rarer than pseudoprimes for the base 2. The counts are taken from the article by Pomerance, Selfridge, and Wagstaff referenced at the end of this chapter.

n	# of ps-primes < n	# of strong ps-primes < n
10^3	3	0
10^6	245	46
10^9	5 597	1 282
25×10^9	21 853	4 842

The strong pseudoprime test becomes much more exclusive if we run it on several bases. The first pseudoprime for bases 2, 3, and 5 is 1729, and there are 2522 such pseudoprimes less than 25×10^9. The first strong pseudoprime for bases 2, 3, and 5 is 25 326 001, and there are only 13 such strong pseudoprimes less than 25×10^9. If we run our test on bases 2, 3, 5, and 7, then there are 1770 pseudoprimes less than 25×10^9 and exactly one strong pseudoprime

$$32150\ 31751 = 151 \times 751 \times 28351.$$

This then gives us a solid primality test for integers below 25×10^9. If n passes the strong pseudoprime tests for bases 2, 3, 5, and 7, if n is less than 25×10^9, and if n is not equal to 32150 31751 then n is prime. Of course, we are still in the range where we can use trial division to prove primality.

There is good news regarding a strong pseudoprime analog of the Carmichael numbers. There is none. We will be proving that every composite number fails the strong pseudoprime test for at least a quarter of the bases less than itself. In fact, it has been shown that if n is composite then it must fail the strong pseudoprime test for at least half the bases less than n. In theory, this gives us a test for primality: If n passes the strong pseudoprime test for more than half the bases less than n, then n is prime. Of course this test would be much slower than trial division and so is impractical. But it does demonstrate that we can have very high confidence in the outcome of a series of strong pseudoprime tests.

6.2   Refining Fermat's Observation

To prove that composite n must fail a strong pseudoprime test, it is useful to go back to the beginning of this chapter and ask a simple question. As we saw, if n is prime and $n = 2m + 1$, then n divides exactly one of the two factors:

$$b^m - 1 \text{ or } b^m + 1.$$

Is there a simple way of telling which one it must divide?

The following theorem is the first step toward answering this question. It was discovered by the same Leonard Euler that we met in Section 3.4.

Theorem 6.2 (Euler's Criterion) *Let p be an odd prime: $p = 2m + 1$, and let b be a positive integer not divisible by p. Then*

$$b^m \equiv 1 \,(\mathrm{mod}p),$$

if and only if there exists an integer t such that

$$b \equiv t^2 \,(\mathrm{mod}p).$$

Proof (one direction): We will just do half of the proof for now. We need another result before finishing it. Assume that such a t exists. Then

$$b^m \equiv t^{2m} \equiv t^{p-1} \,(\mathrm{mod}\ p).$$

But t cannot be divisible by p and so we can use Theorem 3.2,

$$b^m \equiv t^{p-1} \equiv 1 \,(\mathrm{mod}\ p).$$

Q.E.D.

This motivates the following definition.

Definition: Given an integer n and a prime modulus p, if n is relatively prime to p and if there exists an integer t such that

$$n \equiv t^2 \,(\mathrm{mod}\ p),$$

then we say that n is a *quadratic residue modulo p*.

We can find all quadratic residues mod p just by squaring each of the positive integers less than p and looking at their residues. Thus for $p = 7$ we have

$$
\begin{aligned}
1^2 &\equiv 1 \,(\mathrm{mod}\ 7), & 2^2 &\equiv 4 \,(\mathrm{mod}\ 7), & 3^2 &\equiv 2 \,(\mathrm{mod}\ 7), \\
4^2 &\equiv 2 \,(\mathrm{mod}\ 7), & 5^2 &\equiv 4 \,(\mathrm{mod}\ 7), & 6^2 &\equiv 1 \,(\mathrm{mod}\ 7).
\end{aligned}
$$

So the quadratic residues mod 7 are 1, 2, and 4.

Since $i^2 = (-i)^2$, the number of quadratic residues is at most half of $p - 1$. Exercise 6.8 asks you to prove that if

$$i^2 \equiv j^2 \,(\text{mod } p), \text{then}$$
$$i \equiv j \text{ or } -j \,(\text{mod } p),$$

so that each quadratic residue comes up exactly twice. This means that the number of quadratic residues which are positive and less than p is $(p-1)/2$.

The next theorem was observed and conjectured by John Wilson (1741-1793) and proved by Joseph-Louis Lagrange (1736-1813) in 1773.

Theorem 6.3 (Wilson's Theorem) : *The integer n divides $(n-1)! + 1$ if and only if n is prime.*

Proof. We leave it as an exercise to prove that if n is composite, then either $n = 4$ and 4 does not divide $3! + 1 = 7$, or n divides $(n-1)!$ and so does not divide $(n-1)! + 1$.

Assume therefore than n is prime. For each positive integer i less than n, Lemma 4.1 tells us that there exists a unique positive integer j less than n such that

$$i \times j \equiv 1 \,(\text{mod } n).$$

Also, if i is not 1 or $n-1$ then i and j must be distinct because otherwise

$$i^2 \equiv 1 \,(\text{mod } n),$$

which means that the prime n divides

$$i^2 - 1 = (i-1) \times (i+1).$$

Thus when we take the product $(n-1)!$, we can pair up the integers from 2 to $n-2$ with their inverses modulo n:

$$(n-1)! \equiv 1 \times (\text{product of 1's}) \times (n-1) \,(\text{mod } n)$$
$$\equiv n - 1 \equiv -1 \,(\text{mod } n).$$

And so n divides $(n-1)! + 1$.

$$\text{Q.E.D.}$$

This also is a primality test but an impractical one. We can now finish the proof of Theorem 6.2.

Proof of Theorem 6.2 (other direction): Assume that b is not a quadratic residue modulo p. We need to show that

$$b^m \equiv -1 \,(\text{mod } p).$$

For each positive integer i less than p, there exists a unique positive integer j less than p such that

$$i \times j \equiv b \,(\text{mod } p).$$

To find j, first find the j' such that $i \times j' \equiv 1(\text{mod } p)$ and then multiply j' by b to obtain j. If

$$i \times j \equiv i \times k \equiv b \,(\text{mod } p),$$

then multiplying each of the first two quantities by the inverse of i modulo p tells us that

$$j \equiv k \,(\text{mod } p),$$

and thus j is unique modulo p. Since b is not a quadratic residue, i can never equal j.

We take the product of the positive integers less than p, and pair them up into pairs whose product is b modulo p.

$$
\begin{aligned}
(p-1)! &\equiv\ b \times b \times \cdots \times b \,(\text{mod } p) \\
&\equiv\ b^m \,(\text{mod } p),
\end{aligned}
$$

but by Theorem 6.3

$$(p-1)! \equiv -1 \,(\text{mod } p).$$

Therefore p divides $b^m + 1$, and so it does not divide $b^m - 1$.

Q.E.D.

6.3 No "Strong" Carmichael Numbers

We will conclude this chapter with a proof that if n has at least two distinct prime divisors, then there is a base for which n fails the strong pseudoprime test. The case where n is a power of a prime will be proved in Chapter 9, Theorem 9.13.

Lemma 6.4 : *Let r and s be positive integers with $g = \gcd(r, s)$. Let p be a prime and b an integer not divisible by p. If*

$$b^r \equiv 1 \,(\mathrm{mod}\,p) \quad and \quad b^s \equiv 1 \,(\mathrm{mod}\,p), then$$
$$b^g \equiv 1 \,(\mathrm{mod}\,p).$$

Proof. Let $r = m \times g$, $s = n \times g$, so that m and n are relatively prime. Then p divides

$$b^r - 1 = (b^g - 1) \times (1 + b^g + b^{2g} + \cdots + b^{(m-1)\times g}),$$

and p also divides

$$b^s - 1 = (b^g - 1) \times (1 + b^g + b^{2g} + \cdots + b^{(n-1)\times g}).$$

By Lemma 3.5, the two second factors are relatively prime. The prime p might divide one of them but it cannot divide both. Since it does divide both left-hand sides, it must divide $b^g - 1$.

Q.E.D.

Theorem 6.5 : *Let n be an odd composite number with at least two distinct prime factors, say p and q. Write*

$$p = 2^a \times s + 1,$$
$$q = 2^b \times t + 1,$$

where s and t are odd, a and b are at least 1. Order the primes so that

$$a \leq b.$$

Let m be any integer relatively prime to n such that m is a quadratic residue modulo p and is not a quadratic residue modulo q. (Such integers m exist by the Chinese Remainder Theorem, Theorem 4.4.) Then n will fail the strong pseudoprime test for the base m.

Proof. Write n as

$$n = 2^c \times u + 1,$$

where u is odd and c is at least 1. We know that if n is a pseudoprime base m, then each prime dividing n divides exactly one of the factors

$$m^u - 1, m^u + 1, m^{2u} + 1, \ldots, m^{2^{c-1} \times u} + 1,$$

and n passes the strong pseudoprime test base m if and only if they all divide the same factor. If an odd prime divides

$$m^{2^{j-1} \times u} + 1,$$

then it divides

$$m^{2^j \times u} - 1 = (m^{2^{j-1} \times u} + 1) \times (m^{2^{j-1} \times u} - 1),$$

but it does not divide

$$m^{2^{j-1} \times u} - 1 = (m^{2^{j-1} \times u} + 1) - 2.$$

Let j be the smallest integer such that p divides

$$m^{2^j \times u} - 1,$$

and let k be the smallest integer such that q divides

$$m^{2^k \times u} - 1.$$

The exponents j and k are at least 0 and at most c.

Since m is a quadratic residue modulo p, we know by Theorem 6.2 that p divides

$$m^{2^{a-1} \times s} - 1.$$

By Lemma 6.4, p divides

$$m^{gcd(2^j \times u, 2^{a-1} \times s)} - 1, \quad \text{or equivalently}$$

$$m^{gcd(2^j \times u, 2^{a-1} \times s)} \equiv 1 \pmod{p}.$$

If a were less than or equal to j, then we would have

$$gcd(2^j \times u, 2^{a-1} \times s) = 2^{a-1} \times gcd(u, s),$$

and so $2^{a-1} \times u$ would be a multiple of $gcd(2^j \times u, 2^{a-1} \times s)$. Thus we would have that

$$m^{2^{a-1} \times u} \equiv 1 \pmod{p},$$

contradicting the minimality of j. We have proved that

$$a > j.$$

We now look at q. Our base m is not a quadratic residue modulo q, so by Theorem 6.2 q does not divide

$$m^{2^{b-1} \times t} - 1.$$

It does divide

$$m^{2^b \times t} - 1.$$

Again using Lemma 6.4, q must divide

$$m^{gcd(2^b \times t, 2^k \times u)} - 1.$$

If b is larger than k, then

$$gcd(2^b \times t, 2^k \times u) = 2^k \times gcd(t, u),$$

and so

$$m^{2^{b-1} \times t} \equiv 1 \pmod{q},$$

contradicting the fact that m is not a quadratic residue modulo q. Therefore b must be less than or equal to k. Combining our inequalities, we have that

$$j < a \le b \le k.$$

In particular, this says that j and k cannot be equal. Therefore p and q divide distinct factors and n fails the strong pseudoprime test for the base m.

Q.E.D.

REFERENCE

C. Pomerance, J. L. Selfridge, and S. S. Wagstaff , Jr., "The pseudoprimes to 25×10^9", *Math. of Computation*, **35**(1980), 1003-1026.

6.4   EXERCISES

6.1 Write a program to implement Algorithm 6.1 that will run the strong pseudoprime test for each of the bases 2, 3, 5, and 7.

6.2 Use Theorem 6.2 to write a program that uses modular exponentiation to determine whether or not b is a quadratic residue modulo p.

6.3 For each odd prime p less than 100, determine whether or not 2 is a quadratic residue modulo p. Can you see any patterns?

6.4 Same problem as Exercise 6.2 but with 2 replaced by 3.

6.5 Same problem as Exercise 6.2 but with 2 replaced by 5.

6.6 What are the quadratic residues modulo 101? Can you see any patterns?

6.7 For each pair of odd primes, p and q, which are both less than 50, check whether p is a quadratic residue modulo q and then check whether q is a quadratic residue modulo p. Can you see any patterns?

6.8 Prove that if p is a prime and if

$$i^2 \equiv j^2 \pmod{p}, \quad \text{then}$$
$$i \equiv j \text{ or } -j \pmod{p}.$$

6.9 The number $645 = 3 \times 5 \times 43$ is a pseudoprime base 2 but not a strong pseudoprime for the base 2. Which of the factors on the right-hand side of Equation 6.2 are divided by each of the primes dividing 645?

6.10 Prove that if n is composite and larger than 4, then n divides $(n-1)!$. Consider two cases: n is the square of a prime or $n = a \times b$ where a and b are distinct integers larger than 1.

6.11 What goes wrong if we try to replace the prime p in Theorem 6.2 with a composite number n? Find values of b and odd n which are relatively prime and for which there is a t such that

$$b \equiv t^2 \pmod{n},$$

but

$$b^{(n-1)/2} \not\equiv 1 \pmod{n}.$$

6.12 Find values of b and odd n which are relatively prime and for which

$$b^{(n-1)/2} \equiv 1 \pmod{n},$$

but for which there is no t such that

$$b \equiv t^2 \pmod{n}.$$

6.13 As we have seen in Chapter 3, if b and n are relatively prime then n divides

$$b^{\phi(n)} - 1.$$

Furthermore, we know that $\phi(n)$ is even as long as n is larger than 2. Show that if b and n are relatively prime and there is a t such that

$$b \equiv t^2 \pmod{n},$$

then n divides

$$b^{\phi(n)/2} - 1.$$

6.14 Show that the converse is not necessarily true. The converse is the following statement which depends on your choice of n and which we will call $S(n)$.

$S(n)$: Given any integer b which is relatively prime to n and for which

$$b^{\phi(n)/2} \equiv 1 \pmod{n},$$

there is some integer t satisfying

$$b \equiv t^2 \pmod{n}.$$

Find an integer n for which $S(n)$ is false.

6.15 Sometimes $S(n)$ is true for composite n. Let $n = 18$, $\phi(18) = 6$. Prove that $S(18)$ is true.

6.16 Given an n for which $\phi(n)$ is even but not divisible by 4, let k equal $(\phi(n) + 2)/4$. Prove that if

$$
\begin{aligned}
b^{\phi(n)/2} &\equiv 1 \pmod{n}, \text{ then} \\
b &\equiv (b^k)^2 \pmod{n}.
\end{aligned}
$$

Thus prove that $S(n)$ is true whenever 4 does not divide $\phi(n)$.

6.17 Prove or find a counterexample: $S(n)$ is true if and only if 4 does not divide $\phi(n)$.

6.18 Let $\mathrm{PROD}(n)$ be the product of the positive integers less than and relatively prime to n. Prove that if $\mathrm{PROD}(n)$ is not congruent to 1 modulo n then $S(n)$ is true.

6.19 Prove or find a counterexample: $S(n)$ is true if and only if PROD(n) is not congruent to 1 modulo n.

6.20 For which composite n up to 30 is $S(n)$ true? Can you make any guesses about when $S(n)$ is true?

6.21 Let p, q be distinct odd primes dividing n. Show that exactly one-quarter of the positive integers less than and relatively prime to n are quadratic residues modulo p and are not quadratic residues modulo q.

7

Quadratic Reciprocity

> "The general theorem to which we have given the name of *reciprocity law between two primes* (is) the most remarkable and fertile in the theory of numbers."
> – Adrien-Marie Legendre

7.1 The Legendre Symbol

The problem of finding an efficient algorithm for deciding when an integer is a quadratic residue for a given prime is one that occupied the attention of the greatest mathematicians of the late 18th and early 19th centuries. As we have seen, it was Euler who translated the problem of finding the residue of $b^{(p-1)/2}$ into determining whether or not b is a quadratic residue modulo p. Adrien-Marie Legendre (1752-1833) invented the commonly used notation for working on this problem.

Definition: Let p be an odd prime and n an integer. The *Legendre symbol* (n/p) is defined to be 0 if p divides n, $+1$ if n is a quadratic residue modulo p and -1 otherwise.

One of the nice consequences of this definition is that Theorem 6.2 can be restated in the following form.

Corollary 7.1 *If p is an odd prime and n is any integer, then*

$$n^{(p-1)/2} \equiv (n/p) \,(\mathrm{mod}\, p).$$

Setting $n = -1$ in Corollary 7.1 gives us a very useful result.

Corollary 7.2 *The Legendre symbol $(-1/p)$ is $+1$ if $p \equiv 1 \,(\mathrm{mod}\, 4)$ and is -1 if $p \equiv 3 \,(\mathrm{mod}\, 4)$.*

Corollary 7.1 also implies the following properties of the Legendre symbol.

Corollary 7.3 *Given any integers a and b*

$$(a \times b/p) = (a/p) \times (b/p).$$

If $a \equiv b \pmod{p}$, then

$$(a/p) = (b/p).$$

If p does not divide a then

$$(a^2/p) = 1.$$

The next result, known as Gauss' Criterion, was discovered by Carl Friedrich Gauss (1777-1855), the mathematician who made the greatest contributions to the solution of the problem of deciding when an integer is a quadratic residue. The criterion is not terribly practical in itself, but will lead to an efficient algorithm.

Theorem 7.4 (Gauss' Criterion) *Let p be an odd prime and b a positive integer not divisible by p. For each positive odd integer $2i - 1$ less than p, let r_i be the residue of $b \times (2i - 1)$ modulo p:*

$$r_i \equiv b \times (2i - 1) \pmod{p}, \quad 0 < r_i < p.$$

Let t be the number of r_i which are even. Then

$$(b/p) = (-1)^t.$$

Example: Let $p = 7$, then $i = 1$, 2, or 3. If $b = 2$ (which we know to be a quadratic residue) then

$$r_1 \equiv 2 \times 1 \equiv 2 \pmod{7}, \quad r_2 \equiv 2 \times 3 \equiv 6 \pmod{7}, \quad r_3 \equiv 2 \times 5 \equiv 3 \pmod{7},$$

so that $t = 2$ and $(2/7) = +1$. If $b = 3$ (which we know is not a quadratic residue modulo 7) then

$$r_1 \equiv 3 \times 1 \equiv 3 \pmod{7}, \quad r_2 \equiv 3 \times 3 \equiv 2 \pmod{7}, \quad r_3 \equiv 3 \times 5 \equiv 1 \pmod{7},$$

so that $t = 1$ and $(3/7) = -1$.

Proof. Write $p = 2m + 1$. There are m positive odd integers less than p. Relabel the residues so that $r_1, r_2, \ldots, r_t$ are all even and $r_{t+1}, r_{t+2}, \ldots, r_m$ are all odd. Let $a_1, a_2, \ldots, a_m$ be the positive odd integers less that p ordered so that

$$r_i \equiv b \times a_i \pmod{p}.$$

Consider the integers $p - r_1, p - r_2, \ldots, p - r_t, r_{r+1}, r_{t+2}, \ldots, r_m$. These are all positive odd integers less than p. We claim that they are all distinct. Since the a_i's are all distinct modulo p, there are no repetitions among the first t nor among the last $m - t$. It suffices to prove that we cannot have

$$p - r_i = r_j,$$

where i is at most t and j is larger than t. If we did, then we would have that

$$
\begin{aligned}
p = r_i + r_j &\equiv b \times a_i + b \times a_j \,(\text{mod } p) \\
&\equiv b \times (a_i + a_j) \,(\text{mod } p).
\end{aligned}
$$

Since p does not divide b, it must divide $a_i + a_j$. But $a_i + a_j$ is even (because they are each odd) and

$$0 < a_i + a_j < 2p.$$

And so p cannot divide $a_i + a_j$.

Since $p - r_1, \ldots, p - r_t, r_{t+1}, \ldots, r_m$ are m distinct odd positive integers less than p, they must be all of them, and thus

$$
\begin{aligned}
a_1 &\times a_2 \times \cdots \times a_m \\
&= (p - r_1) \times (p - r_2) \times \cdots \times (p - r_t) \times r_{t+1} \times r_{t+2} \times \cdots \times r_m \\
&\equiv (-1)^t \times r_1 \times r_2 \times \cdots \times r_m \,(\text{mod } p) \\
&\equiv (-1)^t \times b \times a_1 \times b \times a_2 \times \cdots \times b \times a_m \,(\text{mod } p) \\
&\equiv (-1)^t \times b^m \times a_1 \times a_2 \times \cdots \times a_m \,(\text{mod } p).
\end{aligned}
$$

We can divide both sides by the product of the a_i's since they are all relatively prime to p. Therefore:

$$(-1)^t \times b^m \equiv 1 \,(\text{mod } p), \text{or}$$

$$b^m \equiv (-1)^t \,(\text{mod } p),$$

and the theorem follows from Corollary 7.1 and the fact that $m = (p-1)/2$.

$$\text{Q.E.D.}$$

7.2   The Legendre symbol for small bases

For small values of b, this test can be practical. For example, you may have guessed from the data gathered in Exercise 6.3 that $b = 2$ is a quadratic residue when $p \equiv 1$ or $-1 \pmod 8$ and it is not a quadratic residue when $p \equiv 3$ or $-3 \pmod 8$.

This can now be easily proved from Theorem 7.4 since t is even if $p \equiv 1$ or $-1 \pmod 8$, and t is odd if $p \equiv 3$ or $-3 \pmod 8$. As an example, if $p = 8m + 1$, then

$$2 \times 1, 2 \times 3, \ldots, 2 \times (4m - 1)$$

all have even residues modulo p, while

$$2 \times (4m + 1), 2 \times (4m + 3), \ldots, 2 \times (8m - 1)$$

all have odd residues. So the number of even residues is $2m$ which is even. We leave the remaining three cases as an exercise.

Corollary 7.5 *If p is an odd prime then*

$$(2/p) = (-1)^{(p^2 - 1)/8}.$$

Proof: Just verify that $(p^2 - 1)/8$ is even if $p \equiv 1$ or $-1 \pmod 8$ and odd if $p \equiv 3$ or $-3 \pmod 8$.

$$\text{Q.E.D.}$$

If $b = 3$, then t is even if $p \equiv 1$ or $-1 \pmod{12}$ and t is odd if $p \equiv 5$ or $-5 \pmod{12}$. Again we will just do the first case and leave the remaining three cases as an exercise. If $p = 12m + 1$, then

$$3 \times 1, 3 \times 3, \cdots, 3 \times (4m - 1), \text{ and}$$

$$3 \times (8m + 1), 3 \times (8m + 3), \cdots, 3 \times (12m - 1)$$

all have odd residues modulo p, while

$$3 \times (4m + 1), 3 \times (4m + 3), \cdots, 3 \times (8m - 1)$$

all have even residues. So the number of even residues is $2m$ which is even.

Corollary 7.6 *If p is an odd prime, then*

$$
\begin{aligned}
(3/p) &= 1 &&\text{if} && p \equiv 1 && \text{or} && -1 \pmod{12}, \\
&= -1 &&\text{if} && p \equiv 5 && \text{or} && -5 \pmod{12}.
\end{aligned}
$$

We can continue proving corollaries like these indefinitely, but both the proofs and the statements of the corollaries quickly become much more complicated. It was the 18th-century analysis of data such as that generated in Exercise 6.7 that led to a really efficient algorithm for computing the Legendre symbol.

7.3   Quadratic Reciprocity

From Corollary 7.3 we see that if we can compute (p/q) for any pair of distinct odd primes p and q, then we can compute (n/q) for any n whose factorization into primes is known. The data in Exercise 6.7 suggested a relationship between (p/q) and (q/p) to Euler and Legendre. Both Euler and Legendre unsuccessfully tried to prove this relationship. The proof was finally discovered in 1796 by the nineteen-year-old Gauss.

Theorem 7.7 (Quadratic Reciprocity) *If p and q are odd primes and at least one of them is $\equiv 1 \pmod 4$, then*

$$(p/q) = (q/p).$$

If both p and q are $\equiv 3 \pmod 4$, then

$$(p/q) = -(q/p).$$

Example: We can combine this theorem with Corollaries 7.2, 7.3, and 7.5. We give three examples

$$
\begin{aligned}
(5/7) \quad &= \quad (7/5) = (2/5) = -1, \quad \text{so} \\
&\qquad \text{5 is not a quadratic residue mod 7,} \\
&\qquad \text{or equivalently 7 divides } 5^3 + 1.
\end{aligned}
$$

$$
\begin{aligned}
(11/23) \quad &= \quad -(23/11) = -(1/11) = -1, \quad \text{so} \\
&\qquad \text{11 is not a quadratic residue mod 23,} \\
&\qquad \text{or equivalently 23 divides } 11^{11} + 1.
\end{aligned}
$$

$$
\begin{aligned}
(1003/1151) \quad &= \quad ((17 \times 59)/1151) = (17/1151) \times (59/1151) \\
&= \quad (1151/17) \times (-1) \times (1151/59) \\
&= \quad -(12/17) \times (30/59) \\
&= \quad -(4/17) \times (3/17) \times (2/59) \times (3/59) \times (5/59) \\
&= \quad -(3/17) \times (-1) \times (3/59) \times (5/59) \\
&= \quad -(17/3) \times (-1) \times (-1) \times (59/3) \times (59/5) \\
&= \quad -(2/3) \times (2/3) \times (4/5) \\
&= \quad -1, \quad \text{so} \\
&\qquad \text{1003 is not a quadratic residue mod 1151,} \\
&\qquad \text{or equivalently 1151 divides } 1003^{575} + 1.
\end{aligned}
$$

Proof. The theorem is trivially true if $p = q$, so we can assume they are distinct primes and so each Legendre symbol is either $+1$ or -1. As in the statement of Theorem 7.4, let s be the number of positive odd integers less than p of the form $2i - 1$ such that if

$$r_i \equiv q \times (2i - 1) \ \mathrm{MOD}\ p$$

then r_i is even. Let t be the number of positive odd integers less than q of the form $2j - 1$ such that if

$$r_j' = p \times (2j - 1) \text{ MOD } q$$

then r_j' is even. By Theorem 7.4, we know that

$$(p/q) \times (q/p) = (-1)^{s+t}.$$

The theorem will follow if we can show that $s + t$ is even when at least one of our two primes is $\equiv 1 \pmod 4$, and $s + t$ is odd if both primes are $\equiv 3 \pmod 4$.

Consider the set S of all integers of the form $q \times a - p \times a'$ where a runs over the positive odd integers less than p and a' runs over the positive odd integers less than q. As an example, if $p = 5$ and $q = 7$, then a is 1 or 3, and a' is 1, 3, or 5. The set S consists of

$$
\begin{array}{rcl}
7 \times 1 - 5 \times 1 & = & 2, \\
7 \times 1 - 5 \times 3 & = & -8, \\
7 \times 1 - 5 \times 5 & = & -18, \\
7 \times 3 - 5 \times 3 & = & 6, \quad \text{and} \\
7 \times 3 - 5 \times 5 & = & -4.
\end{array}
$$

We leave it as Exercise 7.5 to verify that the numbers this generates are always even, non-zero, and distinct.

Consider those pairs (a, a') for which

$$r = q \times a - p \times a'$$

is positive and less than p. This means that

$$q \times a = p \times a' + r \equiv r \pmod p,$$

and so r is one of the residues counted by s. Furthermore, every residue counted by s arises in this way, for if

$$q \times a \equiv r \pmod p,$$

where a is positive, odd, and less than p and r is positive, even, and less than p, then

$$q \times a - r = p \times a',$$

where a' is positive, odd, and

$$p \times a' < q \times p - r, \quad \text{and therefore}$$

$$a' < q.$$

Similarly, every pair (a, a') for which $q \times a - p \times a'$ is negative and larger than $-q$ corresponds to a residue r,

$$q \times a - p \times a' = -r,$$

$$p \times a' \equiv r \pmod{q},$$

counted by t. And every residue counted by t corresponds to a pair (a, a') for which $q \times a - p \times a'$ is negative and larger than $-q$. Therefore $s + t$ equals the number of elements of S which lie between $-q$ and p.

The proof of the theorem boils down to showing that the number of elements of S between $-q$ and p is even except when p and q are both $\equiv 3 \pmod{4}$.

Let $q \times a - p \times a'$ be an element of S in the desired range. Let

$$\begin{aligned} b &= p - 1 - a \quad \text{and} \\ b' &= q - 1 - a'. \end{aligned}$$

We have that $q \times b - p \times b'$ is also an element of S and

$$q \times b - p \times b' = -q + p - (q \times a - p \times a'),$$

so that

$$-q = -q + p - p < q \times b - p \times b' < -q + p - (-q) = p.$$

Therefore, $q \times b - p \times b'$ is also in the desired range. This means that we can pair up the elements of S in the desired range,

$$(a, a') \longleftrightarrow (b, b'),$$

and so the number of such elements must be even unless some elements pair with themselves. But if

$$\begin{aligned} a &= p - 1 - a, \quad \text{and} \\ a' &= q - 1 - a', \quad \text{then} \\ a &= (p-1)/2 \quad \text{and} \quad a' = (q-1)/2. \end{aligned}$$

This means that there is at most one element of S in the desired range that pairs with itself, and this element exists if and only if $(p - 1)/2$ and $(q - 1)/2$ are both odd. In other words, the number of elements in S in the desired range is odd if and only if $p \equiv 3 \pmod{4}$ and $q \equiv 3 \pmod{4}$.

Q.E.D.

7.4 The Jacobi Symbol

We are now equipped with a very powerful tool that enables us to rapidly calculate (n/p) whenever factorization is not a problem. However, we will often be working with large numbers where factorization may present difficulties. This obstacle was neatly removed by Carl Gustav Jacob Jacobi (1804-1851).

Definition: Let n be an integer and m any positive odd integer,

$$m = p_1 \times p_2 \times \cdots \times p_r,$$

where the p_i's are odd primes which may be repeated. The *Jacobi symbol* (n/m) has the value

$$(n/m) = (n/p_1) \times (n/p_2) \times \cdots \times (n/p_r),$$

where (n/p_i) is the usual Legendre symbol.

The Jacobi symbol (n/m) does not tell us whether n is a quadratic residue modulo m. It is rather a convenience that enables us to dispose of factorizations, except for pulling out powers of 2 which is easy, in the computation of the Legendre symbol. Note that if m is a prime, then the Legendre and Jacobi symbols are identical. The beauty of the Jacobi symbol is that it satisfies the same computational rules as the Legendre symbol.

Theorem 7.8 *Let m and m' be odd positive integers, then*

(a) $(n/m) \times (n/m') = (n/(m \times m'))$,

(b) $(n/m) \times (n'/m) = ((n \times n')/m)$,

(c) $(n^2/m) = 1 = (n/m^2)$, *provided n and m are relatively prime,*

(d) *if $n \equiv n' \,(\mathrm{mod}\,m)$, then $(n/m) = (n'/m)$,*

(e) $(-1/m) = 1$ *if $m \equiv 1 \,(\mathrm{mod}\,4)$, $= -1$ if $m \equiv -1 \,(\mathrm{mod}\,4)$,*

(f) $(2/m) = 1$ *if $m \equiv 1$ or $-1 \,(\mathrm{mod}\,8)$, $= -1$ if $m \equiv 3$ or $-3 \,(\mathrm{mod}\,8)$,*

(g) $(n/m) = (m/n)$ *if n and/or $m \equiv 1 \,(\mathrm{mod}\,4)$, $= -(m/n)$ if n and $m \equiv 3 \,(\mathrm{mod}\,4)$.*

Example: This theorem implies that, aside from pulling out factors of 2 as they arise, one can proceed with quadratic reciprocity without worrying about whether or not the numerator is prime. The last example in which we computed the Legendre symbol (1003/1151) is now much simpler:

$$
\begin{aligned}
(1003/1151) &= -(1151/1003) \\
&= -(148/1003) = -(4/1003) \times (37/1003) \\
&= -(37/1003) = -(1003/37) \\
&= -(4/37) \\
&= -1.
\end{aligned}
$$

Proof: Part (a) comes from the definition of the Jacobi symbol. Parts (b)–(d) are immediate consequences of the definition of the Jacobi symbol and the fact that these properties hold for the Legendre symbol.

For parts (e) and (g), observe that all of the primes, p_i, in the prime factorization of m are either congruent to 1 or 3 (mod 4). The product of two primes, both $\equiv 3$ (mod 4), is congruent to 1 (mod 4). So if an even number of these primes are $\equiv 3$ (mod 4), then m will be $\equiv 1$ (mod 4). On the other hand, if an odd number of them are $\equiv 3$ (mod 4), then m will be $\equiv 3$ (mod 4).

Now $(-1/m) = (-1/p_1) \times (-1/p_2) \times \cdots \times (-1/p_r)$, and $(-1/p_i)$ is -1 if and only if $p_i \equiv 3$ (mod 4). So

$$
\begin{aligned}
(-1/m) &= (-1)^{(\text{\# of } p_i \text{ which are } \equiv 3 \,(\text{mod } 4))} \\
&= 1 \text{ if } m \equiv 1 \,(\text{mod } 4), \\
&= -1 \text{ if } m \equiv 3 \,(\text{mod } 4).
\end{aligned}
$$

The proof of part (g) is clearly true if m and n share a common prime factor (both sides are 0). Assume that m and n are relatively prime. We need to break both n and m into products of odd primes. If $m = p_1 \times \cdots \times p_r$ and $n = q_1 \times \cdots \times q_s$, then $(n/m) \times (m/n)$ is the product of $(q_i/p_j) \times (p_j/q_i)$ over all pairs (i, j) with $1 \leq i \leq s$, $1 \leq j \leq r$. But $(q_i/p_j) \times (p_j/q_i)$ is 1 except when q_i and p_j are both $\equiv 3$ (mod 4). This means that the number of -1's in this product is the number of p_j's which are $\equiv 3$ (mod 4) times the number of q_i's which are $\equiv 3$ (mod 4). So $(m/n) \times (n/m)$ is $+1$ except when both m and n are both congruent to 3 modulo 4.

Part (f) uses the same idea. The integer m will be $\equiv 1$ or -1 (mod 8) if and only if an even number of the p_i are $\equiv 3$ or -3 (mod 8).

Q.E.D.

7.5   Computing the Legendre Symbol

We are done! Theorem 7.8 can be restated as a very fast and efficient algorithm for deciding when n is a quadratic residue modulo p.

Algorithm 7.9 *Given an integer n and an odd prime p, this algorithm computes the Legendre symbol (n/p).*

```
INITIALIZE:      READ n, p
                 legendre ← 1
                 n ← n MOD p
```

Input an integer n and a prime p, legendre records the value of (n/p) which is initially set to 1. n is reduced modulo p.

```
CHECK_IF_0:      IF n = 0 THEN DO
                     legendre ← 0
                     CALL TERMINATE
```

If p divides n then $(n/p) = 0$.

```
MAKE_POSITIVE:   IF n < 0 THEN DO
                     n ← -1 × n
                     IF p MOD 4 = 3 THEN legendre ← -1
```

If n is negative, then we pull out a factor of -1.

```
QUAD_REC_LOOP:   CALL PULL_TWOS
                 WHILE n > 1 DO
                     IF (n-1) × (p-1) MOD 8 = 4 THEN DO
                         legendre ← -1 × legendre
                     temp ← n
                     n ← p MOD n
                     p ← temp
                     CALL PULL_TWOS
```

PULL_TWOS *pulls out all factors of 2 from* n. *If* n *is still larger than* 1, *we exchange* n *and* p, *reduce the new value of* n *modulo* p, *and then take out all factors of* 2. *This is iterated until* n *is* 1.

TERMINATE: WRITE legendre

PULL_TWOS: count ← 0
                    WHILE n is even DO
                        n ← n/2
                        count ← 1 - count
                    IF count × (p × p - 1) MOD 16 = 8 THEN DO
                        legendre ← -1 × legendre
                    RETURN

count *records the parity of the exponent of* 2 *in* n. *New values of* n *and* legendre *are returned to the caller.*

We now have a quick test of whether p divides

$$b^{(p-1)/2} - 1 \quad \text{or} \quad b^{(p-1)/2} + 1.$$

A natural question at this point is whether this knowledge can be used to strengthen our strong pseudoprime test. Recall that the strong pseudoprime test on n for the base b checks that n divides one of the factors

$$(b^t - 1), (b^t + 1), (b^{2t} + 1), \ldots \quad \text{or} \quad (b^{2^{a-1} \times t} + 1), \quad \text{where} \quad n = 2^a \times t.$$

We can rapidly compute the Jacobi symbol (b/n). If n is in fact prime, then the value of this symbol is -1 if and only if n divides the last of these factors. Unfortunately, Pomerance, Selfridge, and Wagstaff have shown that if n passes the strong pseudoprime test, then it will always divide the right factor.

Have we done all this work for nothing?

No. The theorems we have proved give us some very important insights into the structure of the integers, and our Algorithm 7.9 will be playing a crucial role in the factorization techniques and primality proofs throughout the rest of this book.

7.6 EXERCISES

7.1 Find the values of the following Jacobi symbols:

$$(35/53)$$

$$(68/233)$$

$$(126/509)$$

$$(672/1297)$$

$$(1235/3499).$$

7.2 Finish the proof of Corollary 7.5.

7.3 Finish the proof of Corollary 7.6.

7.4 Find a result analogous to Corollaries 7.5 and 7.6 for determining when $(5/p) = 1$.

7.5 Let p and q be distinct odd primes and consider the integers of the form $q \times a - p \times a'$ where a runs over the positive odd integers less than p and a' runs over the positive odd integers less than q. Prove that the numbers generated in this fashion are even, non-zero, and distinct.

7.6 Write a program to implement Algorithm 7.9. Find the values of

$$(267\,980/14\,647\,621)$$

$$(1\,073\,899/38\,149\,201)$$

$$(63\,829\,163/409\,482\,089)$$

$$(381\,902\,654/14682\,98937)$$

$$(83772\,01726/1\,93727\,62237).$$

Compare running times with the program written for Exercise 6.2.

7.7 Does there exist an integer n such that 1009 divides $n^2 - 150$? Justify your answer.

7.8 For what primes p does there exist an integer x such that p divides $x^2 + 1$.

7.9 Given an integer a and a prime p, show that $-a$ is a quadratic residue if and only if there exist integers x and y such that p, x, and y are pair-wise relatively prime and p divides $x^2 + ay^2$.

7.10 Describe in terms of congruence class all of the odd primes $p = 2m + 1$ such that p divides $10^m - 1$.

7.11 Let m be an odd composite integer and n be relatively prime to m. Show by finding counterexamples that each of the following statements is false:

(a) m divides $n^{(m-1)/2} - (n/m)$,

(b) m divides $n^{\phi(m)/2} - (n/m)$,

(c) if $(n/m) = +1$, then there exists an integer t such that

$$n \equiv t^2 \ (\text{mod } m).$$

7.12 Recall that if n is a perfect square then $(n/p) = 0$ or 1 for any odd prime p. If n is not a perfect square but $(n/p) = 0$ or 1 for $p = 3, 5, 7,$ and 11, then we will call n a *pseudo-square*. Write a program to test whether or not n is a pseudo-square.

7.13 What is the probability that a randomly chosen integer will be a square or a pseudo-square?

7.14 How many perfect squares are there less than or equal to one million? How many pseudo-squares are there less than or equal to one million? Compare the efficacy of the pseudo-square test with our pseudoprime tests.

7.15 Modify Fermat's Algorithm (Algorithm 5.1) to speed things up by incorporating the pseudo-square test.

7.16 Prove that if p is odd then $(p^2 - 1)/8$ is odd if and only if $p \equiv 3$ or -3 (mod 8).

7.17 Prove that if p and q are odd, then $(p - 1) \times (q - 1)/4$ is odd if and only if p and q are both $\equiv 3$ (mod 4).

7.18 Given that m is not divisible by 2 or 3, prove that the Jacobi symbol $(3/m)$ is $+1$ if and only if $m \equiv \pm 1$ (mod 12).

7.19 What are the residue classes, p, modulo 28 for which the Legendre symbol $(7/p)$ equals $+1$? If a composite integer m is in one of these residue classes, does it follow that the Jacobi symbol $(7/m)$ is equal to $+1$?

7.20 In what respect does the quadratic reciprocity algorithm (Algorithm 7.9) resemble the *gcd* algorithm (Algorithm 1.7)?

8

The Quadratic Sieve

> " 'How do you catch lions in the desert?' Answer: 'In
> the desert you have lots of sand and a few lions; so
> you take a sieve and sieve out the sand, and the lions
> remain.' "
> – Sir Arthur S. Eddington

8.1   Dixon's Algorithm

Once you have used trial division to get all the small divisors out of your
number, a pseudoprime or strong pseudoprime test has shown that it is
still composite, and you feel that you have exhausted the possibilities of
Pollard's rho and $p - 1$ algorithms, then you need to consider one of the
big guns: the Elliptic Curve Method (ECM), the Continued Fraction Algorithm (CFRAC), or the Quadratic Sieve (QS). The Quadratic Sieve has
now totally supplanted the Continued Fraction Algorithm. As currently
implemented, it is faster than CFRAC for any n of at least 18–20 digits,
the smallest size for which QS is readily usable. The Elliptic Curve Method
has the advantage of being practical from the point where trial division becomes impossible until well into the range where QS can be implemented,
at least 25–30 digits. I will be describing CFRAC in Chapter 11 and ECM
in Chapter 14.

Recall from Section 5.2 that Kraitchik suggested that if we can find
"random" integers x and y such that

$$x^2 \equiv y^2 \pmod{n},$$

then we have a reasonable chance that $gcd(n, x - y)$ is a non-trivial factor
of n. To explain how the Quadratic Sieve finds such integers, it is easiest to
begin with a slightly simpler algorithm suggested by John Dixon in 1981.

We choose a random integer r and compute

$$g(r) = r \times r \ \text{MOD} \ n.$$

Now we factor $g(r)$. We are going to need lots of factored numbers, so we
do not try too hard. We just do trial division up to 10 000, and if that
does not work then we pick a different r. We keep doing this until we have

more $g(r)$'s that are completely factored than primes below the limit of trial division. In this case we would need more than 1229 values of r for which we can factor $g(r)$.

Let $p_1, p_2, \ldots, p_{1229}$ be the first 1229 primes. If $g(r)$ factors completely then we can write it as

$$g(r) = p_1^{a_1} \times p_2^{a_2} \times \cdots \times p_{1229}^{a_{1229}},$$

where most of the a_i's will be zero. The factorization of $g(r)$ can be recorded by the vector

$$v(r) = (a_1, a_2, \ldots, a_{1229}).$$

If all of the entries of $v(r)$ are even, then $g(r)$ is a perfect square and we have solved our problem because

$$g(r) \equiv r^2 \pmod{n}.$$

Unfortunately, this is a very unlikely occurence. But we have lots of these $v(r)$'s, more than the length of the vector. That means that we can find a sum of distinct $v(r)$'s which does have all even entries. This is accomplished by setting

$$w(r) = (b_1, b_2, \ldots, b_{1229}), \quad \text{where}$$

$$
\begin{aligned}
b_i &= 0, \quad \text{if} \quad a_i \quad \text{is even} \\
&= 1, \quad \text{if} \quad a_i \quad \text{is odd}.
\end{aligned}
$$

We then do Gaussian elimination modulo 2 on the resulting vectors to find a subset of the r's for which the sum of the corresponding $v(r)$'s has all even coordinates.

Since the sum of these selected $v(r)$'s is a vector with even entries, the product of the corresponding $g(r)$'s will be a perfect square. We get a congruence that looks like

$$g(r_1) \times g(r_2) \times \cdots \times g(r_t) \equiv r_1^2 \times r_2^2 \times \cdots \times r_t^2 \pmod{n},$$

both sides of which are perfect squares. We now have at least a 50–50 chance that this yields a factor of n. If not, we go back and find a different subset of the r's for which the $w(r)$'s are linearly dependent modulo 2.

This is a probabilistic factorization technique. We have no guarantee that it will *ever* give us a factor of n, at least in any specified length of time like the age of the universe. But in practice it will spit out a monster factor faster than any deterministic algorithm.

Gaussian elimination, especially when all the entries are 0 or 1, goes very fast, even with a 1230 by 1230 matrix. What takes time is finding those

1230 completely factored $g(r)$'s because most r's chosen at random will not yield a $g(r)$, all of whose prime factors are less than 10 000.

It is not unreasonable to treat the values of $g(r)$ as random numbers less than n, which means that most of them are roughly the same size as n (see Exercise 8.1). The largest prime factor of $g(r)$ is expected to be about $g(r)^{0.63}$. We can only factor $g(r)$ if the largest prime factor is less than 10 000. If n, and thus $g(r)$, is a 25-digit integer, we cannot factor it unless the largest prime divisor is roughly $g(r)^{0.16}$. The probability of that happening is only about 1/50000, which means that we need to choose around 62 million values for r in order to get 1230 $g(r)$'s that we can factor.

Actually, the situation is not quite as dismal as this. If we choose our r's to be close to the square root of n then the $g(r)$'s will be close to $2\sqrt{n}$. But it will still take almost a million of them to factor a 25-digit integer. This is where the sieve enters.

8.2   Pomerance's Improvement

Carl Pomerance's idea, first proposed in 1981, is to incorporate in this procedure a sieve, like the sieve of Eratosthenes described in Algorithm 2.3, which enables us to simultaneously do the trial division on a million numbers without doing any division.

Instead of choosing the r's at random, let

$$k = \lfloor \sqrt{n} \rfloor,$$

where $\lfloor a \rfloor$ denotes the greatest integer less than or equal to a. It is read as the *floor of a*. Take for the values of r: $k+1, k+2, \ldots$. If we define $f(r)$ to be

$$f(r) = r \times r - n,$$

then $f(r) = g(r)$ as long as r lies between k and the square root of $2n$.

We want to find the $f(r)$'s that factor into primes less than 10 000. Let p be any odd prime less than 10 000. We assume we have already done trial division on n up to 10 000 so we know that p does not divide n. If p divides $f(r)$, then

$$n \equiv r^2 \pmod{p};$$

the Legendre symbol (n/p) must be $+1$. This means that we only need to consider those primes less than 10 000 for which $(n/p) = +1$, in other words about half of them. The set of primes which we try to divide into the $f(r)$'s is called the *factor base*.

If n is a quadratic residue modulo p, then it is the square of one of two residues modulo p:

$$n \equiv t^2 \quad \text{or} \quad (-t)^2 \, (\text{mod } p),$$

which means that r is congruent to either t or $-t$ modulo p. More importantly, if r is congruent to t or $-t$ modulo p, then p must divide $f(r)$.

We can now make two passes down our list of values of $f(r)$. Once we find the first r congruent to t modulo p, we know that $f(r)$ and every p^{th} $f(r)$ thereafter is divisible by p. Then we find the first r congruent to $-t$ modulo p, and again run down the line. Since we *know* which $f(r)$'s are divisible by p without actually doing any division, we can just store the logarithm of $f(r)$ (to single precision) and subtract off the logarithm of p from the appropriate terms. When the remaining logarithm is sufficiently close to 0, we have found an r for which $f(r)$ factors.

Of course, a given prime p may divide into $f(r)$ more than once, so that we should also solve the congruences

$$x^2 \equiv n \, (\text{mod } p^a), \tag{8.1}$$

where p is an odd prime and where the exponent on p runs up to about

$$\frac{2 \log L}{\log p},$$

where L is the largest prime in the factor base.

In practice, however, this means that we will be doing the most sieving on the small primes. There are two problems with sieving over the small primes. First of all, it is slow going. When $p = 3$ we are subtracting $\log 3$ from every third entry. When $p = 311$, a single sieving run takes less than $1/100^{\text{th}}$ of the time because we are subtracting $\log 311$ from every 311th entry. Secondly, $\log 3 = 1.098\ldots$ is much smaller than $\log 311 = 5.739\ldots$.

A solution to this, proposed and implemented by Robert Silverman, is to ignore higher powers, but to be a little more generous on the cut-off for an entry that is accepted as probably completely factorable over the factor base. There will be few enough of these that trial division over the factor base can be used to decide if they really do factor completely.

Even with a million entries, the sieving goes quite quickly. It takes time to solve the quadratic congruence for each prime in the factor base, but the combined time of finding the solutions and doing the sieving is much faster than running trial division a million times.

In summary, there are three steps to the Quadratic Sieve:

1. Find a factor base and solve the congruences

$$x^2 \equiv n \, (\text{mod } p)$$

for each prime p in the factor base.

2. Perform the sieving operation to find sufficient $f(r)$'s which can be completely factored over the factor base.

3. Use Gaussian elimination to find a product of the $f(r)$'s which is a perfect square.

These three steps will be examined in more detail in the next three sections.

8.3   Solving Quadratic Congruences

How large should the factor base be? The factor base should be large enough that there is a reasonable probability that a given $f(r)$ will factor. Balanced against this is the need to keep the factor base small enough that we can do Gaussian elimination on a matrix whose dimensions are the size of the factor base. To help you choose the size of the factor base, I include the following table, condensed from a much more thorough discussion of the size of the kth largest prime divisor of an arbitrary integer which can be found in the Knuth and Trabb-Pardo article listed in the references to this chapter.

a	probability that largest prime divisor of n is $< n^{1/a}$
2	3.07×10^{-1}
3	4.86×10^{-2}
4	4.91×10^{-3}
5	3.55×10^{-4}
6	1.96×10^{-5}
7	8.75×10^{-7}
8	3.23×10^{-8}
9	1.02×10^{-9}
10	2.8×10^{-11}

The prime 2 is exceptional and we will dispose of that first. The number n to be factored is, of course, odd. If it is congruent to 3 or 7 modulo 8 then $r^2 - n$ is divisible by 2 when r is odd and it is never divisible by any higher power of two. If n is congruent to 5 modulo 8 then $r^2 - n$ is divisible by 4 when r is odd, but is never divisible by 8. If it is congruent to 1 modulo 8, then $r^2 - n$ is divisible by at least 8 whenever r is odd.

We would clearly like to have n congruent to 1 modulo 8. There is no reason why it cannot be. If the number you have been given to factor is congruent to 5 modulo 8, then multiply it by 3, if congruent to 3 then multiply by 5, and if congruent to 7 then multiply by 7. Remember that the Quadratic Sieve does not find factors of one size faster than those of another, so that running the Quadratic Sieve on $3n$ does not decrease your chances of finding a large factor of n.

This is an example of the use of a *multiplier*, replacing n by a multiple of n in order to increase the odds that $r^2 - n$ will factor completely. More information on multipliers can be found in the literature referenced at the end of this chapter. Remember that if a prime divides your multiplier, then it will divide $r^2 - n$ if and only if it divides r. Also remember to put your multiplier in *before* you find the factor base.

For each odd prime p in the factor base, we need to solve the congruence

$$x^2 \equiv n \pmod{p}, \tag{8.2}$$

where n is a quadratic residue modulo p. The cases where $p \equiv 3 \pmod 4$ or $\equiv 5 \pmod 8$ are given in Theorem 8.1. Theorem 8.2 is valid for any odd prime, but it is slightly slower than the procedures described in Theorem 8.1.

Theorem 8.1 *If n is a quadratic residue modulo the prime p and*

(1) if $p = 4k + 3$, then

$$x \equiv n^{k+1} \pmod{p},$$

is a solution to Equation (8.2).

(2) If $p = 8k + 5$ and $n^{2k+1} \equiv 1 \pmod{p}$, then

$$x \equiv n^{k+1} \pmod{p},$$

is a solution to Equation (8.2).

(3) If $p = 8k + 5$ and $n^{2k+1} \equiv -1 \pmod{p}$, then

$$x \equiv (4n)^{k+1} \times \left(\frac{p+1}{2} \right) \pmod{p},$$

is a solution to Equation (8.2).

Proof. Since n is a quadratic residue mod p, we know that $n^{(p-1)/2} \equiv 1 \pmod{p}$. If $p = 4k + 3$ then

$$(n^{k+1})^2 = n^{2k+2} = n \times n^{(p-1)/2} \equiv n \pmod{p}.$$

If $p = 8k + 5$, then

$$n^{4k+2} \equiv 1 \pmod{p},$$

which implies that

$$n^{2k+1} \equiv 1 \quad \text{or} \quad -1 \, (\text{mod } p).$$

If it is congruent to $+1$ then

$$(n^{k+1})^2 = n^{2k+1} \times n \equiv n \, (\text{mod } p).$$

If it is congruent to -1 then

$$
\begin{aligned}
(4n)^{2k+2}/4 &= 2^{4k+2} \times n^{2k+2} \\
&\equiv -1 \times (-n) \, (\text{mod } p),
\end{aligned}
$$

because 2 is not a quadratic residue modulo p.

Q.E.D.

Theorem 8.2 *Let n be a quadratic residue modulo an odd prime p and let h be chosen so that the Legendre symbol $(h^2 - 4n/p)$ is -1. Define a sequence $v_1, v_2, \ldots$ by the recursion*

$$
\begin{aligned}
v_1 &= h, \\
v_2 &= h^2 - 2n, \\
v_i &= h \times v_{i-1} - n \times v_{i-2}.
\end{aligned}
$$

We then have that

$$
\begin{aligned}
v_{2i} &= v_i^2 - 2n^i, \text{ and} \\
v_{2i+1} &= v_i \times v_{i+1} - h \times n^i,
\end{aligned}
$$

and a solution of Equation (8.2) is given by

$$x \equiv v_{(p+1)/2} \times \left(\frac{p+1}{2} \right) (\text{mod } p).$$

This algorithm was suggested by D. H. Lehmer in 1969. Its justification will have to await our study of the properties of continued fractions and Lucas sequences in Chapter 12. The proof will be given in Section 12.4. For now, you will have to accept its validity on faith. However, for any given problem it is easy to check that the specific output satisfies your quadratic congruence and I recommend verifying that it does before you proceed with the Quadratic Sieve. Note that a satisfactory h can easily be found by randomly testing different values. Each value you test has a 50% chance of passing, so this should not take long.

The relationship between v_{2i} and v_i can be used to compute an arbitrary v_j in approximately $\log j$ steps in much the same way that Algorithm 3.3 exponentiates in time proportional to the logarithm of the exponent. Knowing v_i and v_{i+1}, we can compute v_{2i}, v_{2i+1}, and v_{2i+2}. Whether we keep v_{2i} and v_{2i+1} or v_{2i+1} and v_{2i+2} depends on the binary expansion of j. We make this explicit in the following algorithm.

Algorithm 8.3 *This algorithm computes v_j modulo p as defined in Theorem 8.2. We assume that h has already been found. We input n, h, j and p and then will only keep track of $v = v_k$, $w = v_{k+1}$ and $m = n^k$.*

```
INITIALIZE:     READ n, h, j, p
                m ← n
                v ← h
                w ← (h × h - 2 × m) mod p
                CALL BINARY(j)
```

n *is known to be a quadratic residue* mod p. h *is chosen so that* h^2 - 4n *is not a quadratic residue* mod p. j *is a positive integer. The last line converts* j *to binary notation.*

```
COMPUTE_LOOP:   FOR k = t - 1 to 1 BY -1 DO
                    x ← (v × w - h × m) MOD p
                    v ← (v × v - 2 × m) MOD p
                    w ← (w × w - 2 × n × m) MOD p
                    m ← m × m MOD p
                    IF bₖ = 0 THEN w ← x
                    ELSE DO
                        v ← x
                        m ← n × m MOD p
```

If v *is* v_k *and* w *is* v_{k+1}, *then the new value of* v *is* v_{2k}, *the new value of* w *is* v_{2k+2}, *and the new value of* x *is* v_{2k+1}. m *keeps track of the power of* n *modulo* p.

```
TERMINATE:      WRITE v
```

```
BINARY(j):        i ← 0
                  WHILE j > 0 DO
                      i ← i + 1
                      bᵢ ← j MOD 2
                      j ← ⌊j/2⌋
                  t ← i
                  RETURN
```

Return values of t *and* bᵢ *to caller.*

8.4   Sieving

One of the first implementations of the Quadratic Sieve was by Gerver at Rutgers in 1982 on a 47-digit integer. The first step, solving the congruences, took seven minutes. The third step, the Gaussian elimination took approximately six minutes. But it needed roughly seventy hours of CPU time to do the sieving. The moral is that this is the place to look for ways of speeding things up.

It is easiest to explain the techniques of the sieving process with an example. Let us factor $n = 499\,94860\,12441$. It is worth observing that this number factors very quickly by either Pollard rho or Pollard $p - 1$. In general, you do not use the Quadratic Sieve on a thirteen digit number. But it will serve for purposes of illustration. We will be looking for a factor base of thirty primes and will start with ten thousand values for r.

The floor of the square root of $499\,94860\,12441$ is $2\,235\,953$. Rather than taking the ten thousand integers r satisfying

$$2\,235\,953 < r < 2\,245\,954,$$

it makes life a little easier to take values of r which straddle the square root

$$2\,230\,953 < r < 2\,240\,954,$$

$$r = 2\,230\,953 + i \,, \ 1 \leq i \leq 10\,000.$$

This keeps the value of $f(r) = r^2 - n$ closer to 0, and the smaller $f(r)$ is, the better are our chances that it will factor. Of course, we do now get negative values of $f(r)$, but we can factor a -1 out of them and treat -1 as the first of our thirty primes in the factor base.

For each prime p in the factor base, n must be a quadratic residue modulo p. Since n is congruent to 1 modulo 8, we can treat 8 as an element of the factor base dividing $r^2 - n$ whenever r is odd. We now run up our list of primes until we find 28 more primes satisfying

$$(n/p) = +1.$$

In this case they are given by

3	19	59	163	229	277	359
5	31	61	181	241	311	367
7	43	67	193	263	331	389
17	47	107	197	271	349	397 .

For each of these primes, we need to solve the congruence

$$n \equiv t^2 \pmod{p}.$$

We give the smallest positive solution t in the position of the corresponding prime. Recall that $p - t$ is another solution.

1	1	14	38	7	39	171
1	5	16	19	18	39	125
2	19	6	86	101	65	69
3	18	32	14	22	52	50

We are now ready to set up the sieve. Rather than calculating the logarithm of the absolute value of $f(2230953+i)$ ten thousand times, Silverman has suggested starting with a vector of zeros to which we add $\log p$ when p divides the corresponding $f(r)$. If we are sieving over $2M$ values, then the logarithm of the absolute value of $(\lfloor \sqrt{n} \rfloor - M + i)^2 - n$ will be approximately

$$\text{TARGET} = (\log n)/2 + \log M.$$

When the sieving is done, there will be few enough entries close to TARGET that we can run trial division over the factor base on them to see exactly which ones factor completely.

How close is close enough? If the remaining unfactored portion is less than the square of the largest prime in the factor base, then it is prime. Even though we may not have a complete factorization over our factor base, we do get a complete factorization which, as we will see later, can still be used. Silverman's suggestion is therefore to set

$$\text{CLOSENUF} = \text{TARGET} - T \times \log(pmax)$$

where $pmax$ is the largest prime in the factor base and T is a constant near 2. For thirteen digit numbers, $T = 1.5$ works nicely.

This Silverman modification does mean that we will miss a few values of r for which $r^2 - n$ factors completely, but it more than compensates in speeding up the sieve.

Returning now to our example, r is odd when i is even, so the first run is to add $\log 8$ to the second entry and every second entry after that. Since 2 230 953 is divisible by 3, we add $\log 3$ to the first entry and then every third and to the second entry and then every third.

Modulo 5, r is congruent to $3 + i$, so that r is congruent to 1 or 4 if and only if i is congruent to 3 or 1 modulo 5. For all such i, we add $\log 5$ to the corresponding entry. Similarly r is congruent to 2 or 5 modulo 7 if and only if i is congruent to 5 or 1. We add $\log 7$ to each of those entries.

We continue in this manner. In general, if SQRT is the floor of the square root of n and we are sieving over an interval of length $2M$, then

$$r = \text{SQRT} - M + i.$$

Given an odd prime p and a solution t between 0 and p of the congruence (8.2), then the first value of i to which we add $\log p$ is

$$p + (t - (\text{SQRT} - M) \text{ MOD } p).$$

Remember that unless p divides the multiplier of n there are two solutions of Equation (8.2), t and $p - t$.

The sieve gives us thirty-nine complete factorizations over the factor base:

$$i = \quad 243; \quad f(r) = \quad 5^2 \times 7 \times 17 \times 107 \times 241 \times 277$$

$$484; \qquad 2^4 \times 3^3 \times 7 \times 19 \times 31 \times 47 \times 241$$

$$649; \qquad 3 \times 7^2 \times 31 \times 59 \times 197 \times 367$$

$$1548; \qquad 2^5 \times 5 \times 7 \times 19 \times 31 \times 67 \times 349$$

$$1755; \qquad 7 \times 19 \times 31 \times 43 \times 263 \times 311$$

$$2336; \qquad 2^3 \times 3 \times 5 \times 7 \times 193 \times 271^2$$

$$2878; \qquad 2^3 \times 3^2 \times 5 \times 7 \times 17 \times 19 \times 43 \times 271$$

$$2916; \qquad 2^4 \times 5 \times 19 \times 43 \times 359 \times 397$$

$$3218; \qquad 2^7 \times 3 \times 5^2 \times 7 \times 17 \times 19 \times 367$$

$$3292; \qquad 2^8 \times 3^3 \times 17 \times 181 \times 359$$

$$3394; \qquad 2^5 \times 3 \times 17 \times 43 \times 263 \times 389$$

$$4094; \qquad 2^3 \times 3^2 \times 19 \times 67 \times 193 \times 229$$

$$4340; \qquad 2^4 \times 3 \times 17 \times 43 \times 241 \times 349$$

$$4476; \qquad 2^6 \times 5^2 \times 17 \times 277 \times 311$$

$$4630; \qquad 2^3 \times 3 \times 59 \times 67 \times 107 \times 163$$

$$4686; \qquad 2^3 \times 5 \times 17^2 \times 331 \times 367$$

$$4786; \qquad 2^6 \times 3^2 \times 5 \times 7 \times 197 \times 241$$

$$4793; \qquad 3 \times 5^2 \times 7 \times 31 \times 163 \times 349$$

$$4866; \qquad 2^5 \times 5 \times 7 \times 47 \times 59 \times 193$$

$$4901; \qquad 3 \times 5^5 \times 7 \times 17 \times 397$$

$$5038; \qquad 2^3 \times 3^2 \times 5 \times 7 \times 193 \times 349$$

$$5043; \qquad 5^2 \times 17 \times 47 \times 59 \times 163$$

$$5101; \qquad 3^3 \times 5^2 \times 7 \times 19 \times 47 \times 107$$

$$5445; \qquad 17 \times 61^2 \times 163 \times 193$$

$$5506; \qquad 2^5 \times 3^3 \times 5 \times 31 \times 61 \times 277$$

$$5683; \qquad 3 \times 5 \times 17^3 \times 181 \times 229$$

$$5840; \qquad 2^3 \times 3^3 \times 19 \times 43 \times 61 \times 349$$

$$6506; \qquad 2^4 \times 3^2 \times 5 \times 67 \times 359 \times 389$$

$$6550; \qquad 2^3 \times 3^2 \times 7 \times 17^2 \times 181 \times 263$$

$$7780; \qquad 2^4 \times 3 \times 17 \times 19 \times 43 \times 47 \times 397$$

$$8216; \qquad 2^3 \times 3^4 \times 5 \times 7 \times 17 \times 163 \times 229$$

$$8528; \qquad 2^3 \times 3 \times 5 \times 17 \times 67 \times 331 \times 349$$

$$8678; \qquad 2^3 \times 3 \times 5 \times 7 \times 31 \times 43 \times 61 \times 241$$

$$8726; \qquad 2^3 \times 3 \times 5^2 \times 17 \times 43 \times 193 \times 197$$

$$8908; \qquad 2^5 \times 3^3 \times 5 \times 47 \times 277 \times 311$$

$$9226; \qquad 2^4 \times 3 \times 5^2 \times 31^2 \times 47 \times 349$$

$$9378; \qquad 2^5 \times 5 \times 7 \times 17 \times 47 \times 61 \times 359$$

$$9763; \qquad 3^2 \times 5 \times 7^2 \times 17 \times 31 \times 59 \times 311$$

$$9908; \qquad 2^4 \times 3^2 \times 5 \times 19 \times 31 \times 197 \times 263.$$

8.5   Gaussian Elimination

We now use Gaussian elimination to find a product of the $f(r)$'s which is a perfect square. For each i for which we can factor $f(2230953 + i)$ we associate a string of thirty binary digits, each column corresponding to one of the thirty primes in the factor base. The digit is 1 if the corresponding prime appears to an odd power and 0 if it appears to an even power. It is convenient to represent the first prime (-1) by the right-most digit and then read the columns from right to left.

With thirty-nine factored numbers, these thirty-nine strings form a matrix of 0's and 1's with thirty-nine rows and thirty columns. In order to keep track of which combination of the $f(r)$'s gives us a perfect square, we adjoin a 39×39 identity matrix to the right. For our example, we give the first ten rows of the resulting matrix:

```
000000010010000010000000110001    100000000000000000000000000000···
000000000010000000001011010101    010000000000000000000000000000···
001000000000100000010010000101    001000000000000000000000000000···
000010000000000001000011011011    000100000000000000000000000000···
000000100100000000000111010001    000010000000000000000000000000···
000000000000010000000000011111    000001000000000000000000000000···
000000001000000000000101111011    000000100000000000000000000000···
110100000000000000000101001001    000000010000000000000000000000···
001000000000000000000001110111    000000001000000000000000000000···
000100000000001000000000100101    000000000100000000000000000000···
```

Starting with the first column on the left-hand side, we find the first string with a 1 in that column and add it modulo 2 ($0 + 0 = 1 + 1 = 0, 0 + 1 = 1 + 0 = 1$) to each of the succeeding strings with a 1 in that column. We then eliminate this first string with a 1 in the first column. The remaining thirty-eight strings now all have 0 in the first column.

The procedure is repeated by finding the first string with a 1 in the second column, adding it to each succeeding string with a 1 in the second column and then eliminating the first string with a 1 in the second column. This is continued until we obtain a string in which the first thirty digits are all 0's. The last thirty-nine digits of this string tell us which of our thirty-nine original strings were added together to get this string which represents a perfect square.

The first string produced in which the first thirty digits are all 0 is

$$...000000000 \quad 00000000000010001000100000000000000100000.$$

The 1's in columns 13, 17, 21, and 34 tell us that the product of the $f(2230953 + i)$'s for the corresponding i's is a perfect square. Those values of i are 4330, 4786, 5038, and 8726. Multiplying the corresponding prime decompositions gives us

$$(2230953+4340)^2 \times (2230953+4786)^2 \times (2230953+5038)^2 \times (2230953+8726)^2$$

$$\equiv 2^{16} \times 3^6 \times 5^4 \times 7^2 \times 17^2 \times 43^2 \times 193^2 \times 197^2 \times 241^2 \times 349^2$$

$$(\bmod\ 4999486012441).$$

We have our congruence of type

$$x^2 \equiv y^2 \ (\bmod\ n).$$

We now compute x and y modulo n, being careful after each multiplication to reduce modulo n so that we do not overflow our accuracy. Unfortunately, in this case the square root of each side modulo 499 94860 12441 is the same: 249 70012 18533. This is the situation we can expect to come up about 50% of the time, that

$$gcd(x - y, n) = 1 \quad \text{or} \quad n.$$

But all is not lost. We started with thirty-nine factorizations. We can continue the Gaussian elimination looking for another vector where the first thirty digits are all zero. The next one that comes up is

$$\ldots 000000000 \quad 10101000111101001010001001010000000000.$$

We follow the same procedure as before. For x and y we get

$$x = 299\ 99039\ 16061; \quad y = 199\ 95820\ 96380.$$

Unfortunately, the *gcd* of the difference of these numbers and 499 94860 12441 is 1. We continue the Gaussian elimination, next producing

$$\ldots 000000000 \quad 10100001010000111001000011100011000000.$$

This time the values of x and y are

$$x = 366\ 53002\ 35664; \quad y = 91\ 58474\ 68484.$$

The *gcd* of $x - y$ and 499 94860 12441 is 999 961 and we have our factorization:

$$499\ 94860\ 12441 = 999\ 961 \times 4\ 999\ 681.$$

8.6   Large Primes and Multiple Polynomials

Among the many refinements of the Quadratic Sieve which have been suggested and implemented, two have shown themselves to be particularly useful in cutting the computing time: the use of large primes and the use of multiple polynomials.

When we ran the sieve on the example in Section 8.5, there were 138 values of i for which we got as far as trial division. Only 39 of these factored completely over the factor base. The remaining 99 factored as a product of primes from the factor base times an additional factor that was less than $397^{1.5}$ (and thus the extra factor is necessarily prime). The large prime refinement uses these extra factorizations with large primes in them.

There is a reasonably good chance that several large primes will appear more than once in these extra factorizations. In our example, if we sieve over the interval of length 8000, we will turn up 32 values of $f(r)$ that factor completely over the factor base. However, we also obtain three factorizations that involve the large prime 449. This means that the product of any two of these three $f(r)$'s will be a product of primes in the factor base times a perfect square (namely, 449^2). This is as good as a complete factorization over the factor base since our object is to find a product of the $f(r)$'s which is a perfect square.

Of the three possible products, we only keep two as any two of the corresponding 30-digit binary strings will add up to the third. We also have two factorizations that involve 443 and two factorizations that include the factor 1097. Thus, keeping track of large primes adds four factorizations to the 32 $f(r)$'s which factor completely over the factor base. With a total of 36 30-digit vectors, we will get at least six distinct products of $f(r)$'s that are perfect squares.

The second refinement was suggested by Peter Montgomery. The $f(r)$'s are smallest, and thus most likely to factor, when r is close to the square root of n. In our example of the sieve of length 10000, sixteen of the thirty-nine $f(r)$'s which factor completely over the factor base have an r which lies within 1000 of the square root of 499 94860 12441. Montgomery's refinement sieves over a shorter interval but with several different quadratic polynomials in r. Instead of just

$$f(r) = r^2 - n,$$

we look at polynomials of the form

$$F(r) = ar^2 + 2br + c.$$

If $n = b^2 - ac$, then

$$
\begin{aligned}
a \times F(r) &= a^2 r^2 + 2abr + ac \\
&= a^2 r^2 + 2abr + b^2 - n \\
&= (ar + b)^2 - n.
\end{aligned}
$$

As before, if a prime p divides $a \times F(r)$ then n is a quadratic residue modulo p, so that we do not need to change our factor base.

The number to be factored hits its minimum at $r = -b/a$. We want to choose a, b, and c so as to minimize both

$$
-F(-b/a) = n/a
$$

and the extreme values at the edge of the sieved interval

$$
F(-M - b/a) = F(M - b/a) = a \times M^2 - n/a.
$$

If M is prescribed, this is accomplished by setting these values equal, that is a should be about

$$
\frac{\sqrt{2n}}{M}.
$$

If a is chosen to be a prime, then we know how to solve the congruence

$$
x^2 \equiv n \pmod{a}.
$$

We choose b to be a solution of this congruence and c to be

$$
c = (b^2 - n)/a.
$$

The Multiple Polynomial Quadratic Sieve (MPQS) described above has a number of nice features. As shown in Exercise 8.24, the upper bound on the value of $F(r)$ is less than the bound on $f(r)$, so that we have a better chance of factoring our numbers. We can use a much shorter sieving interval. If we do not get enough completely factored $F(r)$'s then we generate a new polynomial and sieve again over our shortened interval. Keeping the interval short increases the chances that a given $F(r)$ will factor. One of the nicest features is that the sieving parallelizes perfectly. With N processors, one can assign a different polynomial to each processor and the algorithm runs N times as fast. This aspect was used to dramatic effect in October 1988 when A. K. Lenstra and M. S. Manasse produced the first factorization of a "difficult" 100-digit integer into a product of two primes, and accomplished the factorization by farming out their polynomials to roughly 400 computers around the world.

I have left unanswered the two big questions: How large should the factor base be? How large should M be? Definitive answers to these questions do not exist because the optimum choices will depend on the machine being used. Because of its pipeline architecture, the CRAY runs most efficiently with an M value that is larger than would be appropriate for most other machines. Silverman has suggested values that appear to work well on a VAX/780 when the Quadratic Sieve is used with the large prime and multiple polynomial refinements. I reproduce his table here to serve as a starting point for your own fine tuning. ($K = 1000$.)

Number of Digits	Factor Base Size	M	T	Median Run Time
24	100	5K	1.5	15 sec
30	200	25K	1.5	80 sec
36	400	25K	1.75	400 sec
42	900	50K	2.0	1 800 sec
48	1200	100K	2.0	8 100 sec
54	2000	250K	2.2	27 600 sec
60	3000	350K	2.4	97 200 sec
66	4500	500K	2.6	360 000 sec

REFERENCES

T. R. Caron and R. D. Silverman, "Parallel Implementation of the Quadratic Sieve," *J. Supercomput.*, **1**(1988), 273-290.

James A. Davis, Diane B. Holdright, and Gustavus J. Simmons, "Status Report on Factoring (at the Scandia National Laboratories)," pp. 183-215 in *Advances in Cryptology*, T. Beth, N. Cot, and I. Ingemarsson, eds., Lecture Notes in Computer Science # 209, Springer, Berlin, 1985.

J. D. Dixon, "Asymptotically fast factorization of integers," *Math. Comput.*, **36**(1981), 255-260.

Joseph L. Gerver, "Factoring Large Numbers with a Quadratic Sieve," *Math. of Comput.*, **41**(1983), 287-294.

D. H. Lehmer, "Computer technology applied to the theory of numbers," pp. 117-151 in *Studies in Number Theory*, W. J. LeVeque, ed., Mathematical Association of America Studies in Mathematics # 6, 1969.

Donald E. Knuth and Luis Trabb-Pardo, "Analysis of a simple factorization algorithm," *Theor. Comput. Sci.* **3**(1976), 321-348.

Carl Pomerance, "Analysis and comparison of some integer factoring algorithms," pp. 89-139 in *Computational Methods in Number Theory, Part I*, H. W. Lenstra, Jr. and R. Tijdeman, eds., Mathematical Centre Tracts # 154, Mathematisch Centrum, Amsterdam, 1982.

Carl Pomerance, "The Quadratic Sieve Factoring Algorithm," pp. 169-182 in *Advances in Cryptology*, T. Beth, N. Cot, and I. Ingemarsson, eds., Lecture Notes in Computer Science # 209, Springer, Berlin, 1985.

Robert D. Silverman, "The Multiple Polynomial Quadratic Sieve," *Math. of Comput.*, **48**(1987), Number 177, 329-339.

8.7 EXERCISES

8.1 Let $n = 10^j$ where j is a fixed positive integer. What is the probability that a positive integer less than n will have exactly j digits? What is the probability that it will have at least $j - 1$ digits? At least $j - 2$ digits?

8.2 Show that if n is a thirteen digit integer and $|r - \sqrt{n}| < 5000$, then $|r^2 - n|$ is less than 3.2×10^{10}.

8.3 Using the table at the beginning of Section 8.3, what is the probability that an integer of size about 3×10^{10} will have its largest prime factor less than 400?

8.4 Continuing Exercise 8.3, explain the discrepancy between the expected number of integers to factor completely over the factor base and the fact that we only get 39 complete factorizations when we sieved over the interval of length 10 000.

8.5 Let n be an integer with k digits. If $|r - \sqrt{n}| < M$, show that $|r^2 - n|$ is usually about $M\sqrt{n}$.

8.6 If the factor base consists of primes less than or equal to F, how many primes do you expect to have in the factor base?

8.7 In the original Quadratic Sieve, approximately how many values of r would we need in order to factor a forty-digit integer using primes less than 500 000 in the factor base?

8.8 Verify that the final value of v in Algorithm 8.7 is in fact the value of v_j. This is equivalent to the following problem: The binary expansion of j is given by

$$j = b_1 \times 2^0 + b_2 \times 2^1 + \cdots + b_t \times 2^{t-1}.$$

Show that the final value of c in the following algorithm is j.

```
INITIALIZE:        READ t, b_i
                   c ← 1
                   d ← 2

COMPUTE_LOOP:      FOR i = t - 1 to 1 BY -1 DO
                       c ← 2 × c
                       d ← 2 × d
                       t ← (c + d)/2
                       IF b_i = 0 THEN d ← t
                       ELSE c ← t

TERMINATE:         WRITE c
```

8.9 Solve each of the following congruences:

$$
\begin{aligned}
x^2 &\equiv 7 \,(\mathrm{mod}\ 143), \\
x^2 &\equiv 31 \,(\mathrm{mod}\ 4987), \\
x^2 &\equiv 3 \,(\mathrm{mod}\ 143\,881), \\
x^2 &\equiv 2 \,(\mathrm{mod}\ 327\,853), \\
x^2 &\equiv 26 \,(\mathrm{mod}\ 5\,631\,013), \\
x^2 &\equiv 17 \,(\mathrm{mod}\ 28\,495\,993).
\end{aligned}
$$

8.10 Find a factor base of size 100 to use in applying the Quadratic Sieve to the integer

$$35419\ 05253\ 35205\ 94597\ 94529.$$

8.11 For each prime p in the factor base found in Exercise 8.10, solve the quadratic congruence

$$x^2 \equiv 35419\ 05253\ 35205\ 94597\ 94529 \,(\mathrm{mod}\ p).$$

8.12 We know that the congruence

$$x^2 \equiv 1 \,(\mathrm{mod}\ 8)$$

holds for any $x \equiv \pm 1 \pmod{4}$. Show that if a is odd and

$$x^2 \equiv a \pmod{2^n}$$

has exactly the solutions $x \equiv \pm t \pmod{2^{n-1}}$ then

$$x^2 \equiv a \pmod{2^{n+1}}$$

has either $x \equiv \pm t \pmod{2^n}$ or $x \equiv \pm(2^{n-1} + t) \pmod{2^n}$ as its only solutions.

8.13 Show that if p is an odd prime which does not divide a and if

$$x^2 \equiv a \pmod{p^n}$$

has two solutions: $x \equiv \pm t \pmod{p^n}$, then

$$x^2 \equiv a \pmod{p^{n+1}}$$

has exactly two solutions: $x \equiv \pm(t + k \times p^n) \pmod{p^{n+1}}$, where k satisfies

$$2k \times t \equiv (a - t^2)/p^n \pmod{p}.$$

8.14 Use the Chinese Remainder Theorem to prove that if m and n are relatively prime, if $x^2 \equiv a \pmod{m}$ has s solutions, and $x^2 \equiv a \pmod{n}$ has t solutions, then $x^2 \equiv a \pmod{m \times n}$ has $s \times t$ solutions.

8.15 Find all solutions to each of the following congruences:

$$\begin{aligned}
x^2 &\equiv 2 \pmod{457^3}, \\
x^2 &\equiv 3 \pmod{37^2 \times 457^2}, \\
x^2 &\equiv 5 \pmod{29^3 \times 53^2}.
\end{aligned}$$

8.16 Let m be an odd integer. Show that if m and a are relatively prime, then we can multiply by the inverse of a and complete the square to rewrite

$$ax^2 + bx + c \equiv 0 \pmod{m}$$

in the form

$$y^2 \equiv d \pmod{m}.$$

Find d as a function of a, b, and c. Find x as a function of a, b, c, and y.

8.17 Continuing Exercise 8.16, what happens if m is even?

8.18 Combine the results of Exercises 8.13–16 to write a program that will find all solutions of

$$ax^2 + bx + c \equiv 0 \,(\mathrm{mod}\ m),$$

provided that m is odd and relatively prime to a.

8.19 Prove that if the Jacobi symbol (n/m) is -1 then n cannot be a quadratic residue modulo m. Does $(n/m) = +1$ imply that n is a quadratic residue modulo m?

8.20 In the Gaussian elimination step of the Quadratic Sieve, why does it make sense to eliminate the columns that correspond to the largest primes first?

8.21 In the large prime refinement of the Quadratic Sieve, if we take $T > 2$ then it is possible that the extra factor will not be prime. Explain why this does not affect the running of this refinement.

8.22 In the multiple polynomial refinement of the Quadratic Sieve, each completely factored $a \times F(r)$ is divisible by the large prime a. Explain why this means that for a given polynomial F, if k of our $F(r)$'s factor completely over the factor base, we only get $k - 1$ usable factorizations.

8.23 For the multiple polynomial refinement, Pomerance has suggested that instead of setting a equal to a prime near $\sqrt{2n}/M$, we set $a = p^2$ where p is a prime near $\sqrt{\sqrt{2n}/M}$. Explain why this makes every factorization of $F(r)$ over the factor base usable.

8.24 With a chosen near $\sqrt{2n}/M$, what is the maximum of the absolute value of $F(r)$ for $-M - b/a < r < M - b/a$? Compare this with the upper bound given in Exercise 8.5.

8.25 Continuing Exercises 8.10 and 8.11, use the Multiple Polynomial Quadratic Sieve to finish the factorization of

$$35419\ 05253\ 35205\ 94597\ 95629.$$

9

Primitive Roots and a Test for Primality

> "Yet what are all such gaieties to me
> Whose thoughts are full of indices and
> surds?
> $x^2 + 7x + 53$
> $= 11/3$."
> – Lewis Carroll (Four Riddles)

9.1 Orders and Primitive Roots

In a typical factorization process we take a big number and crack it into lots of little pieces plus a few big pieces that pass the strong pseudoprime tests. We know in our hearts that those big pieces really are primes, but we would like to have absolute certainty. In this chapter we will be examining the notion of primitive roots which will lead us to several primality tests developed by Edouard Lucas in 1876. Perhaps surprisingly, they are still the most efficient way of proving primality for moderately sized primes. Until a few years ago, they were essentially the only tests to prove primality for the really big primes.

We start by recalling Fermat's Theorem 3.2, that if p is a prime and if b is not divisible by p, then

$$b^{p-1} \equiv 1 \pmod{p}.$$

As we proved in Corollary 7.1, if we also know that b is a quadratic residue (*i.e.*, a perfect square) modulo p, then

$$b^{(p-1)/2} \equiv 1 \pmod{p}.$$

If $p - 1$ is divisible by 4, then we can also ask whether

$$b^{(p-1)/4} \equiv 1 \pmod{p}?$$

It should come as no surprise that this is true if and only if b is a *bi-quadratic residue* (*i.e.*, a perfect fourth power) modulo p.

The natural questions that arise at this point are: Given a prime p and an integer not divisible by p, say b, for what exponents m do we have

$$b^m \equiv 1 \pmod{p}?$$

If d divides $p - 1$ and

$$b^{(p-1)/d} \equiv 1 \pmod{p},$$

does it always follow that b is a perfect d^{th} power modulo p? We can also ask the same kinds of questions when the modulus is not prime. If b is relatively prime to n, when do we have

$$b^m \equiv 1 \pmod{n}?$$

We know that

$$b^{\phi(n)/d} \equiv 1 \pmod{n}$$

does not necessarily imply that b is a perfect d^{th} power (see Exercise 6.14 for the case $d = 2$). For what integers n and d does b have to be a perfect d^{th} power?

Definition: Let b and n be positive, relatively prime integers. Let e be the smallest positive integer satisfying

$$b^e \equiv 1 \pmod{n}.$$

Then e is called the *order of b modulo n*.

Note that by Theorem 3.4, e is always less than or equal to $\phi(n)$. As an example, if $n = 7$ then

1	has order	1,
2	has order	3,
3	has order	6,
4	has order	3,
5	has order	6, and
6	has order	2.

Theorem 9.1 *If b has order e modulo n and if j is a positive integer such that*

$$b^j \equiv 1 \pmod{n},$$

then j is a multiple of e.

Proof: We know that j is at least as big as e. Let r denote the remainder when j is divided by e:

$$j = m \times e + r; \qquad 0 \leq r < e.$$

Then we have that

$$1 \equiv b^j \equiv (b^e)^m \times b^r \equiv b^r \pmod{n}.$$

Since e is the smallest positive power of b which is congruent to 1 and r is less than e, r must be 0, and so j is a multiple of e.

Q.E.D.

Corollary 9.2 *For b relatively prime to n, the order of b modulo n must divide $\phi(n)$.*

Note that in the example given above, $\phi(7) = 6$, and the orders we obtained were 1, 2, 3, and 6. One can ask if all divisors of $\phi(n)$ must appear as orders of some relatively prime number. The answer will be "yes" when n is prime and "sometimes" when it is not.

Definition: If b is relatively prime to n and if the order of b modulo n is $\phi(n)$, then we call b a *primitive root modulo n*.

Observe that for $n = 7$, any integer congruent to 3 or 5 modulo 7 is a primitive root.

9.2 Properties of Primitive Roots

Corollary 9.3 *If g is a primitive root modulo n, then every integer which is relatively prime to n is congruent to g^i for some exponent i between 1 and $\phi(n)$, inclusive.*

Proof: We consider the $\phi(n)$ integers: $g, g^2, g^3, \ldots, g^{\phi(n)}$. No two of them are congruent modulo n for if g^i were congruent to g^j, with i less than j, then dividing both sides of the congruence by g^i would give us

$$g^{j-i} \equiv 1 \pmod{n}.$$

Since $j - i$ is less than $\phi(n)$, this would contradict the fact that $\phi(n)$ is the order of g.

Also, every power of g is relatively prime to n. Since there are only $\phi(n)$ congruence classes which are relatively prime to n, the powers of g must exhaust them.

Q.E.D.

If we have a primitive root, then the next theorem guarantees that we will have elements of all orders which divide $\phi(n)$. Recall that lcm denotes the least common multiple.

Theorem 9.4 *If b has order e modulo n, then the order of b^i is*

$$\frac{lcm(e, i)}{i} = \frac{e}{gcd(e, i)}.$$

Proof: Recall from Exercise 1.10 that

$$lcm(e, i) = \frac{e \times i}{gcd(e, i)}.$$

Since $lcm(e, i)$ is a multiple of e, we have that

$$
\begin{aligned}
(b^i)^{lcm(e,i)/i} &= b^{lcm(e,i)} \\
&\equiv 1 \,(\mathrm{mod}\ n).
\end{aligned}
$$

Therefore the order of b^i divides $lcm(e, i)/i$.

Let us write the order of b^i as f/i where f is a multiple of i. This f is less than or equal to $lcm(e, i)$. We also have that

$$
\begin{aligned}
1 &\equiv (b^i)^{f/i} \,(\mathrm{mod}\ n) \\
&\equiv b^f \,(\mathrm{mod}\ n).
\end{aligned}
$$

Thus f must also be a multiple of e, so it is a common multiple of e and i which means it can only be the least common multiple of e and i.

Q.E.D.

Observe that if g is a primitive root modulo n and if d is any divisor of $\phi(n)$, then $g^{\phi(n)/d}$ has order d. We can strengthen this observation with the following corollary which will go a long way toward answering the question posed in Exercise 6.20.

Corollary 9.5 *If there is a primitive root modulo n, if d divides $\phi(n)$, and if b is relatively prime to n, then*

$$b^{\phi(n)/d} \equiv 1 \,(\mathit{mod}\ n),$$

if and only if b is a perfect d^{th} power modulo n.

Proof. It follows from Euler's Theorem (Theorem 3.4) that if b is a perfect d^{th} power modulo n, then the congruence is satisfied because $b \equiv t^d \pmod{n}$ for some t which is also relatively prime to n, and so

$$b^{\phi(n)/d} \equiv t^{\phi(n)} \equiv 1 \pmod{n}.$$

In the other direction, we know that if g is the primitive root then

$$b \equiv g^i \pmod{n},$$

for some integer i. This implies that

$$
\begin{aligned}
1 &\equiv (g^i)^{\phi(n)/d} \pmod{n} \\
&\equiv g^{\phi(n) \times (i/d)} \pmod{n}.
\end{aligned}
$$

Since g has order $\phi(n)$, i/d must be an integer: $i = d \times k$ for some k, and thus b is a perfect d^{th} power:

$$b \equiv (g^k)^d \pmod{n}.$$

Q.E.D.

9.3 Primitive Roots for Prime Moduli

It is our old friend C. F. Gauss who settled the question of when a modulus has a primitive root. In this section we will prove his theorem that there always is a primitive root for any prime modulus.

Lemma 9.6 *Let $P(x)$ be a polynomial of degree t and let p be a prime. If p does not divide the coefficient of x^t in $P(x)$, then the equation*

$$P(x) \equiv 0 \pmod{p}, \qquad (9.1)$$

has at most t incongruent solutions modulo p.

Proof. Assume that Equation (9.1) has at least t incongruent solutions, say $x_1, x_2, \ldots, x_t$. We can divide $x - x_1$ into $P(x)$ to get

$$P(x) = P_1(x) \times (x - x_1) + r,$$

where p divides r because p divides $P(x_1)$. Note that $P_1(x)$ is a polynomial of degree $t - 1$ in x. Thus

$$P(x) \equiv P_1(x) \times (x - x_1) \, (\text{mod } p).$$

Similarly, we can divide $x - x_2$ into $P_1(x)$. We know that $P_1(x_2)$ is divisible by p because

$$0 \equiv P(x_2) \equiv P_1(x_2) \times (x_2 - x_1) \, (\text{mod } p),$$

and p does not divide $x_2 - x_1$. Thus we can write $P_1(x)$ as

$$
\begin{aligned}
P_1(x) &\equiv P_2(x) \times (x - x_2) \, (\text{mod } p), \quad \text{and so} \\
P(x) &\equiv P_2(x) \times (x - x_1) \times (x - x_2) \, (\text{mod } p).
\end{aligned}
$$

$P_2(x)$ is a polynomial of degree $t - 2$ in x. Continuing in this manner, we get that

$$P(x) = c \times (x - x_1) \times (x - x_2) \times \cdots \times (x - x_t) \, (\text{mod } p),$$

where c is a polynomial of degree $t - t = 0$, i.e., c is the constant coefficient of x^t in $P(x)$. Let a be any solution of Equation (9.1), then

$$0 \equiv P(a) \equiv c \times (a - x_1) \times (a - x_2) \times \cdots \times (a - x_t) \, (\text{mod } p).$$

Since p does not divide c, it must divide one of the binomials $a - x_i$, and that means that a is congruent to x_i modulo p for some i.

$$\text{Q.E.D.}$$

Note that this lemma is *not* true if we take a modulus that is not prime. As an example:

$$x^2 \equiv 1 \, (\text{mod } 8),$$

has four incongruent solutions modulo 8.

The next theorem, due to Gauss, more than answers our question about primitive roots when the modulus is prime. It tells us exactly how many there are.

Theorem 9.7 *Let p be a prime and d a divisor of $p - 1$, then the number of positive integers less than p with order d is $\phi(d)$.*

Observe that this says that modulo p there are $\phi(p - 1)$ primitive roots. Since this number is always at least one, primitive roots will always exist when the modulus is a prime.

Proof: Let d be a divisor of $p - 1$ and for the moment let us assume that we do have an integer a of order d modulo p. Then $a, a^2, a^3, \ldots, a^d \equiv 1$ are all distinct modulo p and they are all solutions to

$$x^d \equiv 1 \ (\mathrm{mod} \ p). \tag{9.2}$$

The powers of a give us d incongruent solutions to this equation, so by Lemma 9.6 they are all of the solutions.

Let b be any other integer of order d modulo p. The integer b must also satisfy Equation (9.2), so b is congruent to some power of a. We now use Theorem 9.4. The only values of i for which a^i has order d are those i which are relatively prime to d. And if i is relatively prime to d, then a^i has order d. The number of positive integers less than or equal to d which are relatively prime to d is $\phi(d)$.

What we have proven so far is that if there are any elements of order d, then there are exactly $\phi(d)$ elements of order d. Let $num(d)$ be the number of elements modulo p which have order d. Then

$$num(d) = 0 \quad \mathrm{or} \quad \phi(d).$$

Can $num(d)$ ever be 0? The following lemma shows us that it cannot.

Lemma 9.8 *The sum of $\phi(d)$ where d ranges over all divisors of n is equal to n.*

Example: $n = 12$,

$$\phi(1) + \phi(2) + \phi(3) + \phi(4) + \phi(6) + \phi(12)$$

$$= 1 + 1 + 2 + 2 + 2 + 4 = 12.$$

Proof: Let d be a divisor of n and consider the set $S(d)$ of positive integers less than or equal to n whose greatest common divisor with n is n/d. That is to say, $S(d)$ is the set of x such that

$$gcd(x, n) = n/d \quad \mathrm{and} \quad 1 \le x \le n.$$

Clearly n/d is the smallest element of this set. All of the elements can be written as $k \times n/d$ where k is a positive integer (because x is a multiple of n/d) less than or equal to d (because x is less than or equal to n) and relatively prime to d (because $gcd(x, n) = n/d$). For each k satisfying these three requirements, $k \times n/d$ is a member of the set. Thus $S(d)$ has $\phi(d)$ elements.

Every positive integer less than or equal to n is in exactly one set $S(d)$ for some divisor d of n. So the sum of all the elements of all these sets is n.

Q.E.D.

End of Proof of Theorem 9.7: Recall that $num(d)$ is the number of positive integers less than p which have order d, where d is a divisor of $p-1$. Since every positive integer less than p has an order which divides $p-1$, the sum of $num(d)$ over all divisors of $p-1$ is $p-1$. Since $num(d)$ is either 0 or $\phi(d)$, and the sum of $\phi(d)$ over all divisors of $p-1$ is also $p-1$, we must have that

$$num(d) = \phi(d),$$

for every divisor d of $p-1$.

Q.E.D.

9.4   A Test for Primality

The number of primitive roots modulo p is thus $\phi(p-1)$. This number is always at least one. Though it can vary quite a bit, $\phi(n)$ is on average $6n/\pi^2$, or roughly two-thirds of n. While this estimate is overly optimistic for integers of the form $p-1$, we can still expect a sizeable fraction of the positive integers less than p to be primitive roots modulo p. To find one, just start picking numbers at random. You should not have to go very far before you get one.

How do you recognize a primitive root when you have it? Its order is going to have to be $p-1$, so just make sure that there is no smaller power of our candidate which is congruent to 1 modulo p. You can do this by trying all the powers less than $p-1$, but a little thought shows that we can short-cut this daunting task.

Let b be a randomly chosen integer larger than 1 and less than p. By Corollary 7.1 it must be a quadratic non-residue (*i.e.*, $(b/p) = -1$) or there is no hope for it to be primitive. Let us assume that we know the factorization of $p-1$:

$$p - 1 = 2^{e_1} \times p_2^{e_2} \times \cdots \times p_r^{e_r}.$$

Since the actual order of b divides $p-1$, if b is not primitive then there is at least one prime, p_i, such that $(p-1)/p_i$ is a multiple of the order of b. This means that if b is not primitive then

$$b^{(p-1)/p_i} \equiv 1 \ (\text{mod } p),$$

for some i, $1 \le i \le r$. If we have checked that b is not a quadratic residue, we only have $r-1$ more powers of b to check in order to confirm that it really is a primitive root.

This simple test for primitive roots can be turned around to give us a primality test. Let n be an integer which has passed some strong pseudoprime tests and which we suspect to be prime. Let b be an arbitrary integer larger than 1 and less than n, and assume that n passes the ordinary pseudoprime test for the base b:

$$b^{n-1} \equiv 1 \pmod{n}.$$

The real order of b divides $n - 1$. If n is not prime, then the real order of b is at most $\phi(n)$ which is strictly less than $n - 1$. Let us assume that we know the factorization of $n - 1$, and that $p_1, p_2, \ldots, p_r$ are the distinct primes dividing $n - 1$. If

$$b^{(n-1)/p_i} \not\equiv 1 \pmod{n},$$

for each i from 1 to r, then the real order of b is $n - 1$ which means that n is a prime (and also that b is a primitive root modulo n).

If any of those powers of b are congruent to 1 modulo n, that either means that n is not prime or b is not a primitive root. If we have confidence in the primality of n, we can try more b's until we hit a primitive root. If we try lots of b's and each time find a power which is congruent to 1, then either we are incredibly unlucky or n really is composite. This is one of Lucas' tests for primality.

We can save ourselves a few calculations in this test. Instead of checking that n is prime or a pseudoprime base b, we check that

$$b^{(n-1)/2} \equiv -1 \pmod{n}.$$

This will show both that n is prime or a pseudoprime base b and that the order of b does not divide $(n - 1)/2$. For the other primes dividing $n - 1$: $p_2, p_3, \ldots, p_r$, we can modestly simplify the calculations by checking that

$$b^{(n-1)/2p_i} \not\equiv -1 \pmod{n}.$$

Brillhart, Lehmer, and Selfridge in 1975 realized that you do not have to find a primitive root in order to prove primality, it is enough that for each p_i which divides $n - 1$ you find a b_i satisfying

$$
\begin{aligned}
b_i^{n-1} &\equiv 1 \pmod{n}, \quad \text{but} \\
b_i^{(n-1)/p_i} &\not\equiv 1 \pmod{n}.
\end{aligned}
$$

To see why this is so, let us assume that the factorization of $n - 1$ is given by

$$n - 1 = p_1^{e_1} \times p_2^{e_2} \times \cdots \times p_r^{e_r},$$

and that we have found such b_i's. For each i, the order of b_i divides $n - 1$ but it does not divide $(n - 1)/p_i$. This means that there is a factor of $p_i^{e_i}$ in the order of b_i. Since the order of b_i divides $\phi(n)$, $p_i^{e_i}$ must divide $\phi(n)$. But since this is true for every i, $n - 1$ must divide $\phi(n)$ which can only happen if $n - 1$ equals $\phi(n)$, and so n is prime.

All of this can now be summarized in the following algorithm.

Algorithm 9.9 *This is a primality test for an integer n for which we know the factorization of $n - 1$. If n is indeed composite, then this test may not terminate. It should only be applied to integers which have passed strong pseudoprime tests. We begin by inputting the number to be tested: n, the primes larger than 2 which divide $n - 1$: $p_2, \ldots, p_r$, and a sequence of candidates to be the primitive root: $b_1, b_2, \ldots$, usually chosen from the small primes starting with 2.*

```
INITIALIZE:    READ n, r, pj, bj
               i ← 1
               CALL CHECK
```

CHECK *verifies that* $b_i^{(n-1)/2}$ MOD n *equals* $n - 1$. *If not, then it finds the next* b_i *that satisfies this condition.*

```
TEST_LOOP:     FOR j = 2 to r DO
                  WHILE MODEXPO(bj,(n-1)/(2 × pj),n) = n - 1 DO
                     i ← i + 1
                     CALL CHECK
                  prime ← 1

TERMINATE:     WRITE prime
```

n *is prime if and only if* prime $= 1$.

```
CHECK:         WHILE MODEXPO(bj,(n-1)/2,n) ≠ n - 1 DO
                  IF MODEXPO(bj,(n-1)/2,n) ≠ 1 THEN DO
                     prime ← 0
                     CALL TERMINATE
                  i ← i + 1
               RETURN
```

Return last value of i *to caller.*

`MODEXPO(a,b,n):`

> *Use Algorithm 3.3 to compute* a^b `MOD n. Return this`
> `value to caller.`

Primality testing with this algorithm can be complicated. The problem comes in factoring $n - 1$. Very often one or more of the factors of $n - 1$ are large integers which pass the strong pseudoprime test but which you actually have to prove to be prime before you can proceed. As an example, to prove that

$$n = 61\,89700\,19642\,69013\,74495\,62111$$

is prime, we factor $n - 1$ and obtain

$$n - 1 = 2 \times 3 \times 5 \times 17 \times 23 \times 89 \times 353 \times 397 \times 683 \times 2113 \times 2931542417.$$

To use this factorization, we need to prove that $29315\,42417$ is prime, and for that we need to factor

$$29315\,42416 = 2^4 \times 11 \times 1913 \times 8707.$$

Starting with a very large n, this process may iterate many times before you get a number all of whose factors are known primes.

9.5 More on Primality Testing

Algorithm 9.9 is especially efficient if the number to be tested is one more than a power of 2. Fermat observed that the following numbers are all prime:

$$
\begin{aligned}
2 &+ 1 = 3, \\
2^2 &+ 1 = 5, \\
2^4 &+ 1 = 17, \\
2^8 &+ 1 = 257, \\
2^{16} &+ 1 = 65537.
\end{aligned}
$$

Unfortunately, this does not last.

$$
\begin{aligned}
2^{32} + 1 &= 42949\,67297 \\
&= 641 \times 6\,700\,417.
\end{aligned}
$$

Nevertheless, it motivates the following definition.

Definition: For k greater than or equal to 0, the kth *Fermat number* is

$$F(k) = 2^{(2^k)} + 1.$$

In 1877, the Jesuit priest and mathematician Fr. Jean François Théophile Pépin (1826-1904) published the following test for whether or not $F(k)$ is prime.

Theorem 9.10 $F(k)$ *is a prime for k larger than 1 if and only if*

$$5^{(F(k)-1)/2} \equiv -1 \ (mod \ F(k)).$$

Proof. Since the only prime dividing $F(k) - 1$ is 2, if the congruence is satisfied then $F(k)$ must be prime. All we have to do is show that when $F(k)$ is prime, 5 is not a quadratic residue modulo $F(k)$.

We know that 2^4 is congruent to 1 modulo 5. Since 2^k is a multiple of 4, we see that

$$F(k) = 2^{(2^k)} + 1 \equiv 2 \ (\text{mod } 5).$$

The Legendre symbol $(5/F(k))$ can be computed using quadratic reciprocity:

$$(5/F(k)) = (F(k)/5) = (2/5) = -1.$$

$$\text{Q.E.D.}$$

What if you cannot factor $n - 1$? The following theorem by Henry Cabourn Pocklington (1870-1952) in 1914 shows that you can manage with something less than full factorization.

Theorem 9.11 *Let $n - 1 = F \times R$ where F has a known factorization*

$$F = p_1^{e_1} \times p_2^{e_2} \times \cdots \times p_r^{e_r},$$

and where R is relatively prime to F and less than the square root of n. If for each i from 1 to r there exists a b_i such that

$$b_i^{n-1} \equiv 1 \ (mod \ n), \quad and$$

$$gcd(b_i^{(n-1)/p_i} - 1, n) = 1,$$

then n is prime.

Proof: Let p be any prime dividing n. For each i let a_i be the order of b_i modulo p. We know that a_i divides $p - 1$. Since,

$$b_i^{n-1} \equiv 1 \pmod{p},$$

a_i also divides $n - 1$. On the other hand,

$$b_i^{(n-1)/p_i} \not\equiv 1 \pmod{p},$$

and so a_i does not divide $(n - 1)/p_i$. Thus $p_i^{e_i}$ divides a_i and so also divides $p - 1$. Since this holds for every i, F divides $p - 1$.

But now this implies that every prime p that divides n is larger than F which itself is larger than the square root of n, and that can only happen if n is prime.

Q.E.D.

What if we cannot even satisfy the conditions of Pocklington's theorem? Edouard Lucas came up with another primality test that depends on being able to factor $p + 1$. We will be seeing that test in Chapter 12. The elliptic curve primality tests in Chapter 14 are based on being able to factor other numbers near p.

9.6   The Rest of Gauss' Theorem

What about a modulus that is not prime? Do primitive roots exist then? Consider the orders of the integers modulo 9:

1	has order	1
2	has order	6
4	has order	3
5	has order	6
7	has order	3
8	has order	2.

The integers 2 and 5 are primitive roots modulo 9. Consider the orders of the integers modulo 12:

1	has order	1
5	has order	2
7	has order	2
11	has order	2.

Since $\phi(12) = 4$, there are no primitive roots modulo 12.

A complete characterization of those moduli which have primitive roots was given by Gauss and is summarized in the following theorem.

Theorem 9.12 *There exists a primitive root for the modulus m if and only if m is 2, 4, a power of an odd prime, or twice a power of an odd prime.*

Before proving this theorem we shall use it to clear up a loose end left dangling from Chapter 6. We shall show that if n is a power of a prime and is not a prime, then there is at least one base b relatively prime to n for which n fails the strong pseudoprime test. More than this, there is a b for which n fails the ordinary pseudoprime test.

Theorem 9.13 *Let $n = p^j$ where p is an odd prime and j is at least 2. Let b be any primitive root modulo n. Then n will fail the ordinary pseudoprime test for the base b.*

Proof: Since b is a primitive root, its order is

$$\phi(n) = p^{j-1} \times (p - 1)$$

which is divisible by p. But $n - 1 = p^j - 1$ is not divisible by p and so is not a multiple of the order of b.

Q.E.D.

The proof of Theorem 9.12 will be done in pieces.

Lemma 9.14 *Primitive roots exist for the moduli 2 and 4 but for no higher power of 2.*

Proof: The number 1 is a primitive root modulo 2 and 3 is a primitive root modulo 4. If x is odd then

$$x^2 \equiv 1 \pmod 8.$$

This implies that there can be no primitive roots modulo 8. It also implies that if k is at least 3, then all of the even powers of x from 2 up to $\phi(2^k) = 2^{k-1}$ are congruent to 1 modulo 8. But there are 2^{k-2} of these even powers and only 2^{k-3} of the congruence classes modulo 2^k are congruent to 1 modulo 8. Therefore the powers of x cannot be distinct modulo 2^k.

Q.E.D.

Lemma 9.15 *If p is an odd prime and if g is a primitive root modulo p but not modulo p^k, then $g + p$ is a primitive root modulo p^k.*

Proof: Let e be the order of g modulo p^k. Since p^k divides $g^e - 1$, p also divides it and so e must be a multiple of the order of g modulo p. That is to say, $p - 1$ divides e. Since g is not a primitive root modulo p^k, e divides but is not equal to $(p - 1) \times p^{k-1}$. This implies that

$$g^{(p-1) \times p^{k-2}} \equiv 1 \pmod{p^k}.$$

Since $g + p$ is also a primitive root modulo p, its order modulo p^k must be divisible by $p - 1$ and must divide $(p - 1) \times p^{k-1}$. It will be enough to show that the order of $g + p$ is not a divisor of $(p - 1) \times p^{k-2}$. We use the binomial theorem to expand:

$$(g + p)^{(p-1) \times p^{k-2}} \equiv$$
$$\equiv g^{(p-1) \times p^{k-2}} + (p - 1) \times p^{k-2} \times g^{(p-1) \times p^{k-2} - 1} \times p \pmod{p^k}$$
$$\equiv 1 - g^{(p-1) \times p^{k-2} - 1} \times p^{k-1} \pmod{p^k}.$$

Since p does not divide any power of g, the right-hand side of this congruence is not congruent to 1 modulo p^k, and so the order of $g + p$ is not a divisor of $(p - 1) \times p^{k-2}$. This means that the order of $g + p$ must be $(p - 1) \times p^{k-1} = \phi(p^k)$.

Q.E.D.

Lemma 9.16 *Let p be an odd prime. If g is odd and a primitive root modulo p^k then it is a primitive root modulo $2p^k$.*

Proof: As in the proof of Lemma 9.15, the order of g modulo $2p^k$ must be divisible by the order of g modulo p^k. But the order of g modulo p^k is $\phi(p^k) = \phi(2p^k)$.

Q.E.D.

Note that in Lemma 9.16, if g is even and a primitive root modulo p^k, then we only need to add p^k to it to make it odd and a primitive root modulo p^k. All that remains to prove Theorem 9.13 is to show that for all other moduli there are no primitive roots. The moduli which we have not yet considered can all be written in the form $m \times n$ where m and n are relatively prime and larger than 2. The fact that a primitive root cannot exist in this case follows from the next and last lemma of the chapter.

Lemma 9.17 *If b, m, and n are pairwise relatively prime, then the order of b modulo $m \times n$ is the least common multiple of the order of b modulo m and the order of b modulo n.*

Proof: The Chinese Remainder Theorem gives us a one-to-one correspondence between residues modulo $m \times n$ and pairs of residues modulo m and n:

$$b \leftrightarrow (b \text{ MOD } m, b \text{ MOD } n).$$

Furthermore, this correspondence preserves multiplication. If b corresponds to (s, t) and c corresponds to (u, v), then $b \times c$ corresponds to $(s \times u, t \times v)$.

Since 1 corresponds to $(1,1)$, the order of b modulo $m \times n$ must be a multiple of the order of b modulo m and of the order of b modulo n. Thus the order modulo $m \times n$ is at least as large as the least common multiple of the orders modulo m and n.

In the other direction, any common multiple of the order of b modulo m and the order of b modulo n must be a multiple of the order of b modulo $m \times n$. So the least common multiple of the orders modulo m and n is at least as large as the order of b modulo $m \times n$.

Q.E.D.

This lemma implies that the order of any element modulo $m \times n$ must divide $lcm(\phi(m), \phi(n))$. As long as m and n are larger than 2, both values of ϕ are even and so

$$lcm(\phi(m), \phi(n)) < \phi(m) \times \phi(n) = \phi(m \times n).$$

REFERENCES

John Brillhart, D. H. Lehmer, and J. L. Selfridge, "New primality criteria and factorizations of $2^m \pm 1$," *Math. of Computation*, **39**(1975), 620-647.

D. H. Lehmer, "Tests for primality by the converse of Fermat's theorem," *Bull. Amer. Math. Soc.*, **33**(1927), 327-340.

Fr. Théophile Pépin, "Sur la formule $2^{2^n} + 1$, *Comptes Rendus hebdomadaires des séances de l'Académie des sciences*, **85**(1877), 329-331.

H. C. Pocklington, "The determination of the prime or composite nature of large numbers by Fermat's theorem," *Proc. Camb. Philo. Soc.*, **18**(1914-1916), 29-30.

9.7 EXERCISES

9.1 If b has order 360 modulo m, what is the order of

$$b^{150}?$$

9.2 If b has order 2329 modulo m, what is the order of

$$b^{51}?$$

9.3 If g is a primitive root modulo 67, what powers of g represent the other primitive roots modulo 67?

9.4 Find a primitive root, say g, modulo 31. For each i, $1 \le i \le 30$, find the smallest positive exponent e_i such that

$$g^{e_i} \equiv i \pmod{31}.$$

9.5 Show that if $gcd(d, p-1) = 1$, then every positive integer less than p is congruent to the dth power of some other integer.

9.6 Let d be a divisor of $p-1$ and let a be an integer with order $(p-1)/d$ modulo p. Show that the congruence

$$x^d \equiv a \pmod{p},$$

has exactly d incongruent solutions.

9.7 Find all primitive roots for the following moduli: 13, 25, 54.

9.8 Find a primitive root for each of the following moduli:

$$42\,641, 557\,761, 4\,855\,681.$$

9.9 For each prime p between 500 and 1000, compute

$$\frac{\phi(p-1)}{p-1}.$$

Why don't these values reflect the expected value of $\phi(n)/n$ of $6/\pi^2$?

9.10 Prove that 77 51681 88161 is prime.

9.11 Is $F(6)$ prime? What about $F(7)$ and $F(8)$?

9.12 Prove that if $2^n + 1$ is prime then n is a power of 2.

9.13 What are the possible orders for an integer modulo 1423 and how many positive integers less than 1423 are there of each order?

9.14 What are the possible orders for an integer modulo 35? How many positive integers less than 35 are there of each order?

9.15 What are the possible orders for an integer modulo 799? How many positive integers less than 799 are there of each order?

9.16 Show that if there is no primitive root modulo n then the order of each element divides $\phi(n)/2$.

9.17 Prove that statement $S(n)$ in Exercise 6.14 holds if and only if there is a primitive root modulo n.

9.18 Does

$$b^{\phi(35)/3} \equiv 1 \pmod{35}$$

imply that b is a perfect cube modulo 35?

9.19 Suppose we know that a modulus n has a primitive root. Discuss the likelihood that n is in fact prime. How is the likelihood affected if we also assume that n is odd?

9.20 In the proof of Theorem 9.11, why do we need

$$gcd(b_i^{(n-1)/p_i} - 1, n) = 1$$

instead of just

$$b_i^{(n-1)/p_i} \not\equiv 1 \pmod{n}?$$

9.21 Find a primitive root for each of the following moduli:

$$37, 37^2, 37^3, 2 \times 37^3.$$

9.22 Pépin's Test (Theorem 9.10) is often stated with the 5 replaced by a 3. Prove that if $F(k)$ is prime, then 3 is not a quadratic residue modulo $F(k)$. Can we replace the 5 with a 7 in Pépin's Test?

10

Continued Fractions

> "(Lord William Brouncker), that Most Nobel Man, after having considered this matter, saw fit to bring this quantity by a method of infinitesimals peculiar to him."
>
> – John Wallis

10.1   Approximating the Square Root of 2

We have been following a single thread that began with the Greek problem of characterizing the perfect numbers and has led through Fermat's observation, Euler's theorem, the problem of determining the value of the Legendre symbol, and finally into understanding the multiplicative structure of modular arithmetic. The thread does not end there, but it is time for us to leave it and return to the ancient Greeks to pick up another.

Our new thread starts with the problem of approximating square roots. As we saw in Chapter 1, if an integer is not a perfect square then its square root is not rational. Nevertheless, the only numbers we can actually compute are rational numbers. The Greeks stumbled upon a fast and accurate way of approximating the square root of 2 which was described by Theon of Smyrna in the second century A.D. and is almost certainly much older.

Algorithm 10.1 *This approximates the square root of 2 to within an error of less than* $1/2n^2$.

```
INITIALIZE:     READ n
                a ← 1
                b ← 1

MYSTERY_LOOP:   WHILE b < n DO
                    b ← a + b
                    a ← 2 × b - a

TERMINATE:      WRITE a/b
```

If n is 5000, then the successive approximations we obtain are

$$\begin{aligned}
1/1 &= 1 \\
3/2 &= 1.5 \\
7/5 &= 1.4 \\
17/12 &= 1.416\,666\,66\ldots \\
41/29 &= 1.413\,793\,10\ldots \\
99/70 &= 1.414\,285\,71\ldots \\
239/169 &= 1.414\,201\,18\ldots \\
577/408 &= 1.414\,215\,68\ldots \\
1393/985 &= 1.414\,213\,19\ldots \\
3363/2378 &= 1.414\,213\,62\ldots \\
8119/5741 &= 1.414\,213\,55\ldots,
\end{aligned}$$

accurate to within $\pm 0.000\,000\,02$.

Algorithm 10.1 has a magical quality to it. It is too simple, not at all like the algorithm for computing square roots that is commonly taught in high school. As we try to understand why it works and how it can be extended to compute other square roots, we will be led to many more algorithms that seem equally magical. This is where we will finally understand Algorithms 2.9 and 8.3.

Again it was Pierre de Fermat who made the crucial observation of what is going on here. The key lies in considering the equation

$$a^2 - 2b^2 = \pm 1. \tag{10.1}$$

If we can find integers a and b which satisfy this equation, then a/b will be a good approximation to $\sqrt{2}$ because

$$a^2 = 2b^2 \pm 1,$$

$$\frac{a}{b} = \sqrt{2 \pm \frac{1}{b^2}}.$$

The integers $a = b = 1$ work. How can we find more? The trick comes in thinking of $a + b\sqrt{2}$ as a new kind of "integer". As we saw in Chapter 1, we can add, subtract, multiply, and even divide these *extended integers*. An important consequence of Theorem 1.2 is that each extended integer has a unique representation:

$$a + b\sqrt{2} = r + s\sqrt{2}$$

implies that $a = r$ and $b = s$. Using the extended integers, the left-hand side of Equation (10.1) factors

$$(a + b\sqrt{2}) \times (a - b\sqrt{2}) = \pm 1.$$

Now pick a positive integer i, any i. If we raise the left-hand side of Equation (10.1) to the ith power, the right-hand side will still be either 1 or -1. In terms of our factorization, this says

$$(a + b\sqrt{2})^i \times (a - b\sqrt{2})^i = (\pm 1)^i = \pm 1. \tag{10.2}$$

But $(a + b\sqrt{2})^i$ is another one of our extended integers, say

$$(a + b\sqrt{2})^i = c + d\sqrt{2}. \tag{10.3}$$

Using the binomial theorem to expand the left-hand side of this equation shows that changing the sign of b will not change c and will simply change the sign of d, and so

$$(a - b\sqrt{2})^i = c - d\sqrt{2}. \tag{10.4}$$

If we substitute these extended integers back into Equation (10.2) we get

$$c^2 - 2d^2 = (c + d\sqrt{2}) \times (c - d\sqrt{2}) = \pm 1,$$

and so c and d are two new solutions of Equation (10.1).

Taking $a = b = 1$, the powers of $1 + \sqrt{2}$ have a familiar ring to them:

$$
\begin{aligned}
(1 + \sqrt{2})^2 &= 3 + 2\sqrt{2} \\
(1 + \sqrt{2})^3 &= 7 + 5\sqrt{2} \\
(1 + \sqrt{2})^4 &= 17 + 12\sqrt{2} \\
(1 + \sqrt{2})^5 &= 41 + 29\sqrt{2}, \ldots .
\end{aligned}
$$

If $a + b\sqrt{2}$ is the ith power of $1 + \sqrt{2}$, then the $(i + 1)^{\text{th}}$ power is

$$(a + b\sqrt{2}) \times (1 + \sqrt{2}) = (a + 2b) + (a + b)\sqrt{2},$$

which is exactly the recursion used in Algorithm 10.1.

Not only have we explained Algorithm 10.1, we also see that there is nothing special about the 2. In order to approximate the square root of n, we need to find integers x and y which satisfy

$$x^2 - n \times y^2 = \pm 1. \tag{10.5}$$

Once we have found one such pair, we can generate infinitely many as described in the next theorem which was probably known to the ancient Greeks. It is worth noting that the right-hand side of Equation (10.5) can be -1 only if -1 is a quadratic residue modulo n.

Theorem 10.2 *Let n be a positive integer which is not a perfect square and for which Equation (10.5) has at least one solution in positive integers a and b. Then Equation (10.5) has infinitely many solutions which can be computed recursively by*

$$
\begin{aligned}
x_1 &= a, \\
y_1 &= b, \\
x_{i+1} &= a \times x_i + n \times b \times y_i, \\
y_{i+1} &= b \times x_i + a \times y_i.
\end{aligned}
$$

Proof: We know that if

$$x_i + y_i\sqrt{n} = (a + b\sqrt{n})^i,$$

then x_i, y_i are solutions of Equation (10.5). We only have to verify the recursion satisfied by these x's and y's:

$$
\begin{aligned}
x_{i+1} + y_{i+1}\sqrt{n} &= (x_i + y_i\sqrt{n}) \times (a + b\sqrt{n}) \\
&= (x_i \times a + y_i \times b \times n) + (x_i \times b + y_i \times a)\sqrt{n}.
\end{aligned}
$$

Since a and b are positive, the x's and y's are strictly increasing and so never repeat.

Q.E.D.

10.2   The Bháscara-Brouncker Algorithm

The equation

$$x^2 - n \times y^2 = 1 \qquad (10.6)$$

will always have a solution in positive integers as long as n is positive and not a perfect square. (This equation was incorrectly attributed by Euler to John Pell (1611-1685) and is still commonly known as Pell's equation.) For $n = 3$, $x = 2$ and $y = 1$ works. For $n = 5$, $x = 9$ and $y = 4$ is the smallest solution. The first solution can be quite large as is the case of $n = 61$ where the first solution is

$$
\begin{aligned}
x &= 17663\,19049, \\
y &= 226\,153\,980,
\end{aligned}
$$

or $n = 109$ where the first solution is

$$x = 15807\,06719\,86249,$$
$$y = 1514\,04244\,55100.$$

How do you find the first solution? Does the algorithm in Theorem 10.2 miss any possible solutions? Exactly how accurate is the resulting approximation to $\sqrt{n}$? We will answer these questions in reverse order.

Theorem 10.3 *If a and b are positive integers satisfying*

$$a^2 - n \times b^2 = \pm 1,$$

then the absolute value of the difference between a/b and $\sqrt{n}$ is exactly

$$\frac{1}{b \times (a + b\sqrt{n})}.$$

Proof: We factor the left-hand side of Equation (10.5) and divide by $a + b\sqrt{n}$:

$$a - b\sqrt{n} = \pm 1/(a + b\sqrt{n}),$$
$$a = b\sqrt{n} \pm 1/(a + b\sqrt{n}),$$
$$\frac{a}{b} = \sqrt{n} \pm \frac{1}{b(a + b\sqrt{n})}.$$

Q.E.D.

Since a is always larger than b, this says that the error will always be less than $1/2b^2$. This is exceptionally good. Approximating the square root of 2 with five digit accuracy means approximating it by a rational number whose denominator is 10 000. The error in this case is just under 0.000 002. Approximating the square root of 2 by 8119 / 5741 gives us an error of less than 0.000 000 02.

We next tackle the second-last question.

Theorem 10.4 *If we start the algorithm of Theorem 10.2 with the positive solution of Equation (10.5) which minimizes the numerical value of $a + b\sqrt{n}$, then that algorithm generates all positive solutions of Equation (10.5).*

Proof: Let $x = r$, $y = s$ and $x = u$, $y = v$ be any two solutions of Equation (10.5) and define p and q by

$$(r + s\sqrt{n}) \times (u - v\sqrt{n}) = p + q\sqrt{n}.$$

Then $x = |p|$, $y = |q|$ is also a solution of Equation (10.5) because

$$(r - s\sqrt{n}) \times (u + v\sqrt{n}) = p - q\sqrt{n},$$

and so

$$
\begin{aligned}
p^2 - n \times q^2 &= (p + q\sqrt{n}) \times (p - q\sqrt{n}) \\
&= (r + s\sqrt{n}) \times (u - v\sqrt{n}) \times (r - s\sqrt{n}) \times (u + v\sqrt{n}) \\
&= (r^2 - n \times s^2) \times (u^2 - n \times v^2) = \pm 1.
\end{aligned}
$$

Now, we define e to be the value of $a^2 - n \times b^2$, $e = \pm 1$. If $x = r$, $y = s$ is a solution of Equation (10.5) which is not generated by the algorithm of Theorem 10.2, then there is a positive integer j for which

$$(a + b\sqrt{n})^j < r + s\sqrt{n} < (a + b\sqrt{n})^{j+1}.$$

Define p and q by

$$p + q\sqrt{n} = (r + s\sqrt{n}) \times (a - b\sqrt{n})^j \times e^j.$$

By what we showed in the first part of this proof, $x = |p|$, $y = |q|$ is a solution of Equation (10.5). But if we multiply through our double inequality by $(a - b\sqrt{n})^j \times e^j$ we get

$$1 < p + q\sqrt{n} < a + b\sqrt{n}.$$

The first of these inequalities guarantees that p and q are both positive (see Exercise 10.5) and the second inequality contradicts the fact that $x = a$, $y = b$ is the positive solution which minimizes $x + y\sqrt{n}$.

Q.E.D.

Fermat realized that the crux of the problem lay in finding the first solution of Equation (10.5) and in 1657 he sent out a challenge problem, to find solutions to Equation (10.6). Rather mischievously, he proposed the values $n = 61$ and $n = 109$ as very small numbers "in order not to give you too much trouble." In response, Lord William Brouncker (1620-1684) and John Wallis (1616-1703) sent back the following algorithm which is still the most efficient way of finding the first solution. Wallis was later to credit Brouncker with the discovery of the algorithm.

It seems likely that Fermat himself had some method of attacking this problem since he was able to single out the two difficult cases of $n = 61$ and 109. But there is an algorithm equivalent to Brouncker's which predates Fermat by 500 years. It was found by the Indian mathematician Bháscara Achárya (ca. 1115-1185). For this reason, I will refer to it as the Bháscara-Brouncker Algorithm.

Algorithm 10.5 (Bháscara-Brouncker) *We input n and* sqrt *= (the floor of the square root of n) into this algorithm. It generates five sequences:*

A_i, B_i, C_i, P_i, and Q_i. The ratio P_i/Q_i gives progressively better approximations to the square root of n and in fact we always have the equality:

$$P_i^2 - n \times Q_i^2 = (-1)^i \times C_i.$$

The first time $C_i = 1$ for positive i, the corresponding values of P_i, Q_i will be the first solution of equation (10.5).

```
INITIALIZE:      READ n
                 sqrt ← ⌊√n⌋
                 A₁ ← sqrt
                 B₀ ← 0; B₁ ← sqrt
                 C₀ ← 1; C₁ ← n - sqrt × sqrt
                 P₀ ← 1; P₁ ← sqrt
                 Q₀ ← 0; Q₁ ← 1
                 i ← 1


MYSTERY_LOOP:    WHILE Cᵢ ≠ 1 DO
                     k ← i - 1
                     j ← i
                     i ← i + 1
                     Aᵢ ← ⌊(sqrt + Bⱼ)/Cⱼ⌋
                     Bᵢ ← Aᵢ × Cⱼ - Bⱼ
                     Cᵢ ← Cₖ + Aᵢ × (Bⱼ - Bᵢ)
                     Pᵢ ← Pₖ + Aᵢ × Pⱼ
                     Qᵢ ← Qₖ + Aᵢ × Qⱼ


TERMINATE:       WRITE Pᵢ, Qᵢ, i
```

As an example, I give you the values up to $i = 11$ of the sequences generated when $n = 13$:

i	A_i	B_i	C_i	P_i	Q_i	
1	3	3	4	3	1	
2	1	1	3	4	1	
3	1	2	3	7	2	
4	1	1	4	11	3	
5	1	3	1	18	5	⟸
6	6	3	4	119	33	
7	1	1	3	137	38	
8	1	2	3	256	71	
9	1	1	4	393	109	
10	1	3	1	649	180	⟸
11	6	3	4	4287	1189	

As promised:

$$18^2 - 13 \times 5^2 = 324 - 325 = -1.$$

The next solution occurs the next time C_i is 1:

$$649^2 - 13 \times 180^2 = 421\,201 - 421\,200 = 1.$$

The remainder of this chapter will be spent justifying the Bháscara-Brouncker algorithm and proving that eventually some C_i will be 1.

10.3   The Bháscara-Brouncker Algorithm Explained

There is a general technique for finding rational approximations to arbitrary numbers. Christian Huygens (1629-1695) seems to have been the first to understand it. To approximate, say, the square root of 13, we apply the Euclidean algorithm to the pair $\sqrt{13}$ and 1.

At first glance, this seems a ridiculous thing to do. The Euclidean algorithm will not terminate because if $\sqrt{13}$ and 1 did have a greatest common divisor, then either that *gcd* is rational which means that $\sqrt{13}$ is rational, or the *gcd* is irrational, which means that 1 is irrational. But we are not looking for a greatest common divisor here.

What we do is to iterate the Euclidean algorithm until the remainder is sufficiently close to zero and then stop, say after k iterations. It follows from Theorem 1.10 that the values of m that we have generated are precisely the m's that would occur if we applied the Euclidean algorithm to the continued fraction

$$m_1 + \cfrac{1}{m_2 + \cfrac{1}{m_3 + \cdots \frac{1}{m_k}}}$$

and the integer 1. Thus, the rational number represented by this continued fraction should be a good approximation to the irrational number with which we started, and we can improve the approximation just by continuing the iteration of the Euclidean algorithm.

Applying this to the square root of 13 we get

$$
\begin{array}{rclcl}
\sqrt{13} = & 3.6055512\ldots & = & 3 \times 1 & + & 0.6055512\ldots \\
& 1 & = & 1 \times 0.6055512\ldots & + & 0.3944487\ldots \\
& 0.6055512\ldots & = & 1 \times 0.3944487\ldots & + & 0.2111025\ldots \\
& 0.3944487\ldots & = & 1 \times 0.2111025\ldots & + & 0.1833461\ldots \\
& 0.2111025\ldots & = & 1 \times 0.1833461\ldots & + & 0.0277563\ldots \\
& 0.1833461\ldots & = & 6 \times 0.0277563\ldots & + & 0.0168079\ldots \\
& 0.0277563\ldots & = & 1 \times 0.0168079\ldots & + & 0.0109484\ldots \\
& 0.0168079\ldots & = & 1 \times 0.0109484\ldots & + & 0.0058594\ldots .
\end{array}
$$

The successive rational approximations are

$$3$$
$$3 \;+\; 1/1 = 4$$
$$3 \;+\; 1/(1+1/1) = 7/2$$
$$3 \;+\; 1/(1+1/(1+1/1)) = 11/13$$
$$3 \;+\; 1/(1+1/(1+1/(1+1/1))) = 18/5$$
$$3 \;+\; 1/(1+1/(1+1/(1+1/(1+1/6)))) = 119/33,$$

and then

$$\frac{137}{38}, \quad \frac{256}{71}, \quad \frac{393}{109}, \quad \frac{649}{180}, \quad \cdots$$

It was Euler who first realized that these are exactly the approximations to the square root of 13 which the Bháscara-Brouncker algorithm yields. In fact for any n the Bháscara-Brouncker algorithm amounts to no more than a calculation of the continued fraction approximations of the square root of n.

Actually, this is quite fortunate. To compute the continued fraction approximations to $\sqrt{n}$ using the Euclidean algorithm, we needed a good rational approximation of $\sqrt{n}$. What the Bháscara-Brouncker algorithm says is that we can get the continued fraction approximations just by starting with the greatest integer less than or equal to $\sqrt{n}$. This can be found by using the algorithm given in Exercise 5.7.

Before stating Euler's theorem which explains the Bháscara-Brouncker algorithm, we need a better notation for continued fractions. From now on, instead of

$$m_1 + \cfrac{1}{m_2 + \cfrac{1}{m_3 + \cdots \cfrac{1}{m_k}}},$$

we will write this as

$$m_1 + \frac{1}{m_2+}\,\frac{1}{m_3+} \cdots \frac{1}{m_k}.$$

If the continued fraction does not have a terminating value, then it represents the corresponding irrational number:

$$\sqrt{13} = 3 + \frac{1}{1+}\,\frac{1}{1+}\,\frac{1}{1+}\,\frac{1}{1+}\,\frac{1}{6+}\,\frac{1}{1+}\,\frac{1}{1+}\,\frac{1}{1+}\,\frac{1}{1+}\,\frac{1}{6+} \cdots.$$

Theorem 10.6 *Let $A_i, B_i, C_i, P_i,$ and Q_i be as defined in Algorithm 10.5 where n is not a perfect square. Then the following properties hold for positive values of i:*

$$P_i^2 - n \times Q_i^2 = (-1)^i \times C_i, \tag{10.7}$$

$$0 < A_i < 2 \times \sqrt{n}, \tag{10.8}$$

$$0 < B_i < \sqrt{n}, \tag{10.9}$$

$$0 < C_i < 2 \times \sqrt{n}, \tag{10.10}$$

$$B_i^2 + C_i \times C_{i-1} = n, \tag{10.11}$$

$$\sqrt{n} = A_1 + \cfrac{1}{A_2 +} \cfrac{1}{A_3 +} \cfrac{1}{A_4 +} \cdots, \tag{10.12}$$

let $E_{i+1} = (\sqrt{n} + B_i)/C_i$, then

$$\sqrt{n} = A_1 + \cfrac{1}{A_2 +} \cfrac{1}{A_3 +} \cdots \cfrac{1}{A_i +} \cfrac{1}{E_{i+1}}, \tag{10.13}$$

$$\frac{P_i}{Q_i} = A_1 + \cfrac{1}{A_2 +} \cfrac{1}{A_3 +} \cdots \cfrac{1}{A_{i-1} +} \cfrac{1}{A_i}, \tag{10.14}$$

$$P_i \times Q_{i-1} - P_{i-1} \times Q_i = (-1)^i, \tag{10.15}$$

$$gcd(P_i, Q_i) = 1. \tag{10.16}$$

Furthermore, the sequence of triples (A_i, B_i, C_i) is eventually periodic. We shall denote this period by $\rho(n)$. (For example, $\rho(13) = 5$.)

Proof: The periodicity follows from Equations (10.8)-(10.10) which imply that there are only finitely many triples which can occur. Eventually a succession of two triples repeats itself. By the recursive definition of these sequences, once two successive triples repeat we have entered a loop.

We first prove Equation (10.11) by induction. Note that it is true when $i = 1$. Assume that it is true for i, we will see that it also holds for $i + 1$.

$$\begin{aligned}
B_{i+1}^2 &+ C_{i+1} \times C_i \\
&= (A_{i+1} \times C_i - B_i) \times B_{i+1} \\
&\quad + [C_{i-1} + A_{i+1} \times (B_i - B_{i+1})] \times C_i \\
&= A_{i+1} \times B_{i+1} \times C_i - B_i \times B_{i+1} + C_{i-1} \times C_i \\
&\quad + A_{i+1} \times B_i \times C_i - A_{i+1} \times B_{i+1} \times C_i \\
&= -B_i \times (A_{i+1} \times C_i - B_i) + C_{i-1} \times C_i \\
&\quad + A_{i+1} \times B_i \times C_i \\
&= B_i^2 + C_{i-1} \times C_i = n.
\end{aligned}$$

This concludes the proof of Equation (10.11).

The next piece that we attack consists of Equations (10.12) and (10.13). We first observe that by the definition of E_i, A_i is the greatest integer less than or equal to E_i. We use Equation (10.11) to obtain the following equation for the E's:

$$
\begin{aligned}
E_i & - 1/E_{i+1} \\
&= \frac{\sqrt{n} + B_{i-1}}{C_{i-1}} - \frac{C_i}{\sqrt{n} + B_i} \\
&= \frac{n + \sqrt{n} \times (B_i + B_{i-1}) + B_i \times B_{i-1} - C_i \times C_{i-1}}{C_{i-1} \times (\sqrt{n} + B_i)} \\
&= \frac{B_i \times B_i + \sqrt{n} \times (B_i + B_{i-1}) + B_i \times B_{i-1}}{C_{i-1} \times (\sqrt{n} + B_i)} \\
&= \frac{B_i + B_{i-1}}{C_{i-1}} \\
&= \frac{A_i \times C_{i-1}}{C_{i-1}}, \quad \text{by the recursion formula for } B_i, \\
&= A_i.
\end{aligned}
$$

We can write this equality as

$$E_i = A_i + 1/E_{i+1}. \tag{10.17}$$

Now we have it because

$$
\begin{aligned}
\sqrt{n} &= E_1 \\
&= A_1 + 1/E_2 \\
&= A_1 + \cfrac{1}{A_2 +} \cfrac{1}{E_3} \\
&= \cdots \\
&= A_1 + \cfrac{1}{A_2 +} \cfrac{1}{A_3 +} \cdots \cfrac{1}{A_i +} \cfrac{1}{E_{i+1}} \\
&= \cdots \\
&= A_1 + \cfrac{1}{A_2 +} \cfrac{1}{A_3 +} \cfrac{1}{A_4 +} \cdots .
\end{aligned}
$$

This concludes the proof of Equations (10.12) and (10.13).

We next look at the inequalities (10.8) - (10.10). By Equation (10.17), we know that

$$1/E_{i+1} = E_i - A_i < 1,$$

which means that

$$E_{i+1} > 1.$$

Since A_{i+1} is the floor of E_{i+1}, we can conclude that

$$A_{i+1} > 0.$$

We now establish the bounds on C_i and B_i by induction. Observe that

$$0 < C_0 = 1 < 2\sqrt{n} \quad \text{and}$$
$$0 < B_1 = \lfloor \sqrt{n} \rfloor < \sqrt{n}.$$

Assume that these same bounds hold for C_i and B_{i+1}. Again using Equation (10.17), we have that

$$
\begin{aligned}
0 < E_{i+1} - A_{i+1} &= 1/E_{i+2} \\
&= \frac{C_{i+1}}{\sqrt{n} + B_{i+1}},
\end{aligned}
$$

And thus C_{i+1} must be larger than zero. If we multiply numerator and denominator of this last fraction by $\sqrt{n} - B_{i+1}$ and then use Equation (10.11) to simplify the denominator, we get that

$$
\begin{aligned}
\frac{1}{E_{i+2}} &= \frac{C_{i+1} \times (\sqrt{n} - B_{i+1})}{n - B_{i+1}^2} \\
&= \frac{\sqrt{n} - B_{i+1}}{C_i}.
\end{aligned}
$$

Note that this gives us another equation that can be used to define E_i:

$$E_i = \frac{C_{i-2}}{\sqrt{n} - B_{i-1}}. \tag{10.18}$$

Replacing i by $i + 3$ in this equation gives us

$$0 < E_{i+3} = \frac{C_{i+1}}{\sqrt{n} - B_{i+2}}.$$

Since C_{i+1} is positive, B_{i+2} must be less than $\sqrt{n}$.

If B_{i+2} were not positive, then by the recursion for B we would have that $A_{i+2} \times C_{i+1} - B_{i+1}$ is less than or equal to zero, and therefore

$$\sqrt{n} > B_{i+1} \geq A_{i+2} \times C_{i+1} \geq C_{i+1}.$$

But if C_{i+1} is less than the square root of n and B_{i+2} is not positive, then

$$1 > E_{i+2} - A_{i+2}$$
$$= \frac{\sqrt{n} - B_{i+2}}{C_{i+1}} \geq \frac{\sqrt{n}}{C_{i+1}} > 1.$$

Since 1 cannot be strictly larger than 1, B_{i+2} must be positive.
We finish the inductive argument by observing that

$$1 < E_{i+2} = \frac{\sqrt{n} + B_{i+1}}{C_{i+1}} < \frac{2\sqrt{n}}{C_{i+1}}, \quad \text{and so}$$

$$C_{i+1} < 2\sqrt{n}.$$

Finally, we have that

$$A_i < E_i = \frac{\sqrt{n} + B_{i-1}}{C_{i-1}} < 2\sqrt{n}.$$

This concludes the proofs of inequalities (10.8)-(10.10).

We next prove Equations (10.14)-(10.16); in fact we will prove more than that. Let $(a_1, a_2, a_3, \ldots)$ be any sequence of positive numbers. They do not have to be integers. We define two sequences; p_i and q_i by

$$p_{-1} = q_0 = 0,$$
$$p_0 = q_{-1} = 1,$$
$$p_{i+1} = p_{i-1} + a_{i+1} \times p_i,$$
$$q_{i+1} = q_{i-1} + a_{i+1} \times q_i.$$

I claim that then

$$p_i \times q_{i-1} - p_{i-1} \times q_i = (-1)^i \quad \text{and}$$

$$\frac{p_i}{q_i} = a_1 + \frac{1}{a_2+} \frac{1}{a_3+} \cdots \frac{1}{a_i}.$$

If the a_i's are integers, then the p_i's and q_i's must be integers. It follows from the first of these equalities that for each i, p_i and q_i are relatively prime.

The first of these equalities is proven by induction. We can easily verify that it is true when $i = 0$. Assume that it is true for some i. Then by our recursion we have that

$$p_{i+1} \times q_i - p_i \times q_{i+1}$$
$$= (p_{i-1} + a_{i+1} \times p_i) \times q_i - p_i \times (q_{i-1} + a_{i+1} \times q_i)$$
$$= p_{i-1} \times q_i - p_i \times q_{i-1}$$
$$= -(-1)^i = (-1)^{i+1}.$$

And so it is also true for $i + 1$. This proves Equations (10.15) and (10.16).

We also prove the second identity by induction, first checking the first two values of i:

$$\frac{p_1}{q_1} = \frac{a_1}{1} = a_1,$$

$$\frac{p_2}{q_2} = \frac{1 + a_1 \times a_2}{a_2} = a_1 + \frac{1}{a_2}.$$

We now proceed by induction. Assume that

$$a_1 + \frac{1}{a_2+} \cdots \frac{1}{a_i} = \frac{p_i}{q_i} = \frac{p_{i-2} + a_i \times p_{i-1}}{q_{i-2} + a_i \times q_{i-2}}.$$

We replace a_i by $a_i + 1/a_{i+1}$

$$a_1 + \frac{1}{a_2+} \quad \cdots \quad \frac{1}{a_i+} \frac{1}{a_{i+1}}$$

$$= \frac{p_{i-2} + (a_i + \frac{1}{a_{i+1}}) \times p_{i-1}}{q_{i-2} + (a_i + \frac{1}{a_{i+1}}) \times q_{i-1}}$$

$$= \frac{p_i + \frac{1}{a_{i+1}} \times p_{i-1}}{q_i + \frac{1}{a_{i+1}} \times q_{i-1}}$$

$$= \frac{a_{i+1} \times p_i + p_{i-1}}{a_{i+1} \times q_i + q_{i-1}}$$

$$= \frac{p_{i+1}}{q_{i+1}}.$$

This concludes the proof of Equation (10.14). The fraction $\frac{p_i}{q_i}$ is called the ith *convergent*.

We are finally ready to prove Equation (10.7). From Equation (10.13) and what we have just proven about continued fractions, we see that

$$\sqrt{n} = \frac{P_{i-2} + (A_i + \frac{1}{E_{i+1}})P_{i-1}}{Q_{i-2} + (A_i + \frac{1}{E_{i+1}})Q_{i-1}} \tag{10.19}$$

$$= \frac{P_i + \frac{1}{E_{i+1}} \times P_{i-1}}{Q_i + \frac{1}{E_{i+1}} \times Q_{i-1}}$$

$$= \frac{P_{i-1} + E_{i+1} \times P_i}{Q_{i-1} + E_{i+1} \times Q_i}.$$

We multiply through by the denominator and rewrite E_{i+1} in terms of $\sqrt{n}$, B_i, and C_i.

$$\sqrt{n} \times (Q_{i-1} + Q_i \times (\sqrt{n} + B_i)/C_i)$$
$$= P_{i-1} + P_i \times (\sqrt{n} + B_i)/C_i,$$

which can be rewritten as

$$\sqrt{n} \times (Q_{i-1} + Q_i \times B_i/C_i) + n \times Q_i/C_i$$
$$= \sqrt{n} \times P_i/C_i + P_{i-1} + B_i \times P_i/C_i.$$

The coefficients of $\sqrt{n}$ must be equal. This gives us two equations:

$$P_i = C_i \times Q_{i-1} + B_i \times Q_i, \tag{10.20}$$

$$n \times Q_i = B_i \times P_i + C_i \times P_{i-1}. \tag{10.21}$$

We now use these equations:

$$
\begin{aligned}
P_i^2 - n \times Q_i^2 &= (C_i \times Q_{i-1} + B_i \times Q_i) \times P_i \\
&\quad -(B_i \times P_i + C_i \times P_{i-1}) \times Q_i \\
&= C_i \times (Q_{i-1} \times P_i - P_{i-1} \times Q_i) \\
&= C_i \times (-1)^i.
\end{aligned}
$$

Q.E.D.

10.4 Solutions Really Exist

We still have not proven that Equation (10.5) has a solution, nor that the smallest solution must occur in the Bháscara-Brouncker Algorithm. This was first done by Joseph-Louis Lagrange.

Theorem 10.7 *If n is a positive integer and not a perfect square, then there exists an integer solution to Equation (10.5).*

Proof. Algorithm 10.5 generates infinitely many distinct pairs (P_i, Q_i) which satisfy Equation (10.7). Since we know that C_i is less than $2\sqrt{n}$ and there are only finitely many integers between $-2\sqrt{n}$ and $2\sqrt{n}$, there has to be at least one integer m between $-2\sqrt{n}$ and $2\sqrt{n}$ for which

$$x^2 - n \times y^2 = m \tag{10.22}$$

has infinitely many solutions.

We sort our solutions of Equation (10.22) according to the residues of x and y modulo m. With infinitely many of them to sort, we can find two that agree in both residues:

$$x_1 \equiv x_2 \pmod{m} \quad \text{and} \quad y_1 \equiv y_2 \pmod{m}.$$

We now define the integers u and v by

$$u + v\sqrt{n} = (x_1 + y_1\sqrt{n}) \times (x_2 - y_2\sqrt{n}).$$

We have that

$$
\begin{aligned}
u^2 - n \times v^2 &= (u + v\sqrt{n}) \times (u - v\sqrt{n}) \\
&= (x_1 + y_1\sqrt{n}) \times (x_2 - y_2\sqrt{n}) \times (x_1 - y_1\sqrt{n}) \times (x_2 + y_2\sqrt{n}) \\
&= (x_1^2 - n \times y_1^2) \times (x_2^2 - n \times y_2^2) = m^2.
\end{aligned}
$$

But we also have that

$$
\begin{aligned}
u &= x_1 \times x_2 - n \times y_1 \times y_2 \\
&\equiv x_1 \times x_1 - n \times y_1 \times y_1 \pmod{m} \\
&\equiv 0 \pmod{m}, \quad \text{and} \\
v &= y_1 \times x_2 - x_1 \times y_2 \equiv 0 \pmod{m},
\end{aligned}
$$

and so u and v are both divisible by m. If we set $a = u/m$ and $b = v/m$ then

$$a^2 - n \times b^2 = 1.$$

Q.E.D.

We finish up by showing not just that the first solution of Equation (10.5) must be generated by Algorithm 10.5, but that in fact *every* good approximation of the square root of n is generated by this algorithm.

Theorem 10.8 *Let n be a positive integer which is not a perfect square and let x and y be positive integers such that*

$$|\sqrt{n} - x/y| < 1/2y^2.$$

Then for some i in Algorithm 10.5,

$$x = P_i \quad and \quad y = Q_i.$$

Recall that Theorem 10.3 guaranteed that every solution of Equation (10.5) gives us an approximation to $\sqrt{n}$ with at least this accuracy, and thus all solutions of Equation (10.5) are generated by the Bháscara-Brouncker algorithm.

Proof: Find the largest i such that Q_i is less than or equal to y. If $Q_i = y$, then $P_i = x$ because both P_i/Q_i and x/y differ from $\sqrt{n}$ by less than $1/2y$. We now consider what happens if

$$Q_i < y < Q_{i+1}.$$

Consider the following system of linear equations in a and b:

$$
\begin{aligned}
a \times Q_i + b \times Q_{i+1} &= y, \\
a \times P_i + b \times P_{i+1} &= x.
\end{aligned}
$$

The determinant is $Q_i \times P_{i+1} - Q_{i+1} \times P_i = 1$ or -1 by Equation (10.15). Thus this system has a unique solution in integers a and b. Since y lies between Q_i and Q_{i+1}, a and b cannot both be positive and a cannot be zero.

By Equation (10.7), the real numbers

$$Q_i \times \sqrt{n} - P_i \quad and$$

$$Q_{i+1} \times \sqrt{n} - P_{i+1},$$

have opposite signs, and so

$$
\begin{aligned}
a \times (Q_i \times \sqrt{n} - P_i) \quad and \\
b \times (Q_{i+1} \times \sqrt{n} - P_{i+1}),
\end{aligned}
$$

must have the same sign if b is not zero. It therefore follows that

$$
\begin{aligned}
|y \times \sqrt{n} - x| &= |a \times (Q_i \times \sqrt{n} - P_i) + b \times (Q_{i+1} \times \sqrt{n} - P_{i+1})| \\
&= |a| \times |Q_i \times \sqrt{n} - P_i| + |b| \times |Q_{i+1} \times \sqrt{n} - P_{i+1}| \\
&\geq |Q_i \times \sqrt{n} - P_i|.
\end{aligned}
$$

Dividing this last inequality by Q_i yields

$$|\sqrt{n} - P_i/Q_i| \leq |y \times \sqrt{n} - x|/Q_i \leq 1/2yQ_i.$$

Using this inequality and the fact that P_i and Q_i are relatively prime, we get that

$$\frac{1}{y \times Q_i} \leq \frac{|y \times P_i - x \times Q_i|}{y \times Q_i} \quad = \quad |P_i/Q_i - x/y|$$

$$\leq \quad |P_i/Q_i - \sqrt{n}| + |\sqrt{n} - x/y|$$

$$\leq \quad \frac{1}{2y \times Q_i} + \frac{1}{2y^2}.$$

Multiplying through by $2y^2 \times Q_i$ gives us

$$2y \quad \leq \quad y + Q_i, \quad \text{or equivalently}$$
$$y \quad \leq \quad Q_i,$$

But this contradicts our original assumption that y lay between Q_i and Q_{i+1}.

Q.E.D.

We have gone to an awful lot of trouble if all we want to do is approximate square roots. A careful re-reading of the proofs of Theorems 10.6-10.8 will reveal that we have actually proved a great deal about rational approximations of *any* irrational number. This is worked out in detail in Exercises 10.20-10.23. Also, as we will see in the next chapter, the structure of these sequences can be used to say a great deal about the integers. These sequences will now play an important role in factoring and determining primality.

REFERENCES

Harold M. Edwards, *Fermat's Last Theorem*, Springer-Verlag, New York, 1977.

André Weil, *Number Theory, An approach through history, from Hammurapi to Legendre*, Birkhauser, Boston, 1984.

10.5   EXERCISES

10.1 Use Exercise 1.4 to prove that if a, b, x, and y are integers and n is not a perfect square, then

$$a + b \times \sqrt{n} = x + y \times \sqrt{n},$$

implies that $a = x$ and $b = y$.

10.2 Prove that if

$$u + v\sqrt{n} = (a + b\sqrt{n}) \times (c + d\sqrt{n}),$$

then

$$u - v\sqrt{n} = (a - b\sqrt{n}) \times (c - d\sqrt{n}).$$

10.3 Prove that if

$$
\begin{aligned}
a^2 - n \times b^2 &= x \quad \text{and} \quad c^2 - n \times d^2 = y, \quad \text{and} \\
u + v\sqrt{n} &= (a + b\sqrt{n}) \times (c + d\sqrt{n}),
\end{aligned}
$$

then

$$u^2 - n \times v^2 = x \times y.$$

10.4 Prove that if

$$x + y \times \sqrt{n} = (a + b \times \sqrt{n})^i,$$

then

$$
\begin{aligned}
x &= [(a + b \times \sqrt{n})^i + (a - b \times \sqrt{n})^i]/2 \quad \text{and} \\
y &= [(a + b \times \sqrt{n})^i - (a - b \times \sqrt{n})^i]/2\sqrt{n}.
\end{aligned}
$$

10.5 In the proof of Theorem 10.4, why does

$$p^2 - n \times q^2 = \pm 1 \quad \text{and} \quad p + q\sqrt{n} > 1$$

imply that p and q are both positive? (*Hint*: Show that $p - q\sqrt{n}$ must lie between -1 and 1.)

10.6 Write a program which uses Algorithm 10.5 to compute square roots to 200 digit accuracy. Compute the square roots of 2, 3, and 5 to this accuracy.

10.7 If we run Algorithm 10.5 with $n = 61$, the first time that $C = 1$ is when $P = 29718$ and $Q = 3805$. Explain why this does not contradict the assertion that the smallest solution of Equation (10.6) with $n = 61$ is $P = 17663\ 19049$ and $Q = 226\ 153\ 980$.

10.8 Find the first solution to Equation (10.6) for each of the following values of n: 433, 613, 1709.

10.9 What happens to Algorithm 10.5 if the greatest integer less than or equal to the square root of n is incorrectly evaluated?

10.10 Run Algorithm 10.5 with $n = 3$ out to $i = 50$. Which P's and Q's are divisible by 3? by 5? by 7? by 11? by 13? Can you make any conjectures?

10.11 Same problem as Exercise 10.10 but with $n = 5$.

10.12 Find the sequences of the A_i's in the continued fraction expansions of the square roots of each of the integers up to twenty which are not perfect squares. Can you make any conjectures? Can you prove any of your conjectures?

10.13 Compute $\rho(n)$ (the period of Algorithm 10.5) for each $n < 100$ which is not a perfect square. Can you make any conjectures?

10.14 Run twenty iterations of the Euclidean algorithm on the pair $\sqrt{2}$ and 1. How does the size of the remainder compare with the size of the error when the corresponding continued fraction is used to approximate $\sqrt{2}$?

10.15 What irrational number is represented by the infinite continued fraction:

$$1 + \frac{1}{1+}\frac{1}{1+}\frac{1}{1+}\frac{1}{1+}\frac{1}{1+}\frac{1}{1+}\cdots ?$$

(*Hint*: It will have to satisfy the equation $x = 1 + 1/x$.) This number is known as the "Golden Ratio", it is the proportion of length to width which the Greeks found most beautiful.

10.16 Compute the first ten terms of P_i and Q_i in the convergents to the Golden Ratio in Exercise 10.15. These sequences were first described by Leonardo of Pisa (1180-1250), popularly known as Fibonacci.

10.17 What irrational number is represented by the infinite continued fraction

$$1 + \frac{1}{2+} \frac{1}{3+} \frac{1}{1+} \frac{1}{2+} \frac{1}{3+} \frac{1}{1+} \cdots ?$$

(*Hint*: It will have to satisfy the equation: $x = 1 + 1/[2 + 1/(3 + 1/x)]$.)

10.18 It follows from the periodicity of Algorithm 10.5 that the continued fraction for any square root is eventually periodic. Prove that if an irrational number has a continued fraction expansion which is eventually periodic, then it must be the root of a quadratic polynomial with integer coefficients.

10.19 Prove that

$$\left\lfloor \frac{\lfloor \sqrt{n} \rfloor + B_i}{C_i} \right\rfloor = \left\lfloor \frac{\sqrt{n} + B_i}{C_i} \right\rfloor .$$

IN EXERCISES (10.20)-(10.23), LET u BE AN ARBITRARY POSITIVE IRRATIONAL NUMBER WHOSE CONTINUED FRACTION EXPANSION IS GIVEN BY

$$u = a_1 + \frac{1}{a_2+} \frac{1}{a_3+} \cdots .$$

LET $\frac{p_i}{q_i}$ BE THE ith CONVERGENT TO u,

$$\frac{p_i}{q_i} = a_1 + \frac{1}{a_2+} \frac{1}{a_3+} \cdots \frac{1}{a_i} .$$

AND DEFINE e_i BY

$$u = a_1 + \frac{1}{a_2+} \cdots \frac{1}{a_{i-1}+} \frac{1}{e_i} ,$$

SO THAT

$$e_{i+1} = \frac{1}{e_i - a_i} \quad \text{AND} \quad a_{i+1} = \lfloor e_{i+1} \rfloor .$$

10.20 Prove that

$$p_i \times q_{i-1} - p_{i-1} \times q_i = (-1)^i ,$$

and therefore $gcd(p_i, q_i) = 1$.

10.21 Prove that

$$u = \frac{p_i + (e_i - a_i) \times p_{i-1}}{q_i + (e_i - a_i) \times q_{i-1}} .$$

10.22 Prove that

$$u - \frac{p_i}{q_i} = \frac{(-1)^{i+1} \times (e_i - a_i)}{q_i \times [q_i + (e_i - a_i) \times q_{i-1}]},$$

and therefore

$$\left| u - \frac{p_i}{q_i} \right| < \frac{1}{q_i^2}$$

10.23 Let x and y be positive, relatively prime integers which satisfy

$$\left| u - \frac{x}{y} \right| < \frac{1}{2y^2}.$$

Show that $\frac{x}{y}$ must be one of the convergents to u : $x = p_i$ and $y = q_i$ for some i.

10.24 Show that if

$$a^2 - n \times b^2 = e = \pm 1, \quad \text{then}$$

$$(a + b\sqrt{n})^{-1} = e \times (a - b\sqrt{n}).$$

In the extended integers of the form $x + y\sqrt{n}$, we call an integer a *unit* if its inverse is also an integer. Show that if n is positive and not a perfect square then there is unit $a + b\sqrt{n}$ (called the *fundamental unit*) such that every unit is of the form

$$\pm(a + b\sqrt{n})^i$$

for some integer i.

10.25 Prove that the definition of a unit given in Exercise 10.24 is equivalent to the definition given in Exercise 1.1.

11

Continued Fractions
Continued, Applications

"Mazes intricate,
Eccentric, interwov'd, yet regular
Then most, when most irregular they
⠀ ⠀ seem."

– John Milton

11.1 CFRAC

We are now ready to explain the Brillhart-Morrison Continued Fraction Algorithm (commonly known as CFRAC) for factoring large numbers. The original idea is actually due to D. H. Lehmer and R. E. Powers in 1931 and it draws much of its inspiration from Legendre who used the continued fraction expansion in a procedure that restricted the congruence classes of possible divisors, but it was put in its present form by John Brillhart and Michael Morrison who published a thorough account of it in 1975.

The basic approach is the same as in the Quadratic Sieve. We want to find solutions to

$$x^2 \equiv y^2 \pmod{n}. \tag{11.1}$$

If n is the number to be factored, the Quadratic Sieve generates a large number of integers of the form m^2 MOD n which can be completely factored using a set of primes in a relatively small factor base. Once the number of completely factored integers exceeds the size of the factor base, we can find a product of them which is a perfect square. This gives us a solution to the congruence (11.1). As we have seen, there is at least a 50-50 chance that $gcd(n, x - y)$ will be a non-trivial divisor of n.

The difference between the Continued Fraction Algorithm and the Quadratic Sieve lies in how we generate the integers m. Recall that in the Quadratic Sieve, we simply take the integers closest to the square root of n. The resulting integer $m^2 - n$ that we need to factor tends to be quite large. Using the Multiple Polynomial variation with an optimal choice of parameters still leaves us with numbers as large as $\frac{M}{\sqrt{2}}\sqrt{n}$ to be factored, and so the probability that it can be factored using the primes in the factor base is fairly small. The Quadratic Sieve gets its speed from the fact that deciding which integers can be factored is accomplished very quickly by means of the sieving technique.

The Continued Fraction Algorithm chooses values of m for which m^2 MOD n has a smaller upper bound, staying in absolute value less than twice the square root of n. The probability that it can be factored using our factor base is thus somewhat higher. The penalty for this is that there is no fast way to decide whether m^2 MOD n can be factored using the factor base. One has to resort to trial division.

The values of m that are used in the Continued Fraction Algorithm are precisely the P_i generated in Algorithm 10.5. Observe that from Equation (10.7) we have that

$$P_i^2 \equiv (-1)^i \times C_i \ (\text{mod } n). \tag{11.2}$$

Equation (10.10) gives us the promised bound on C_i. As in the Quadratic Sieve, we treat -1 as a prime in our factor base. Also note that from Equation (10.7), if a prime p divides C_i, then n is a quadratic residue modulo p. We can thus restrict our factor base to primes p for which $(n/p) = 1$, exactly as we did in the Quadratic Sieve.

We modify Algorithm 10.5 so that it ignores the Q's and only keeps track of P_i MOD n, C_i, and the parity of i. For each new C_i, we call up a trial division subroutine that attempts to factor C_i over the factor base. If it is successful, then we store P_i MOD n, C_i, and the factorization of $(-1)^i \times C_i$. If it is unsuccessful, then we return to Algorithm 10.5 to generate the next set of values. When the number of completely factored C_i's exceeds the size of the factor base, we use Gaussian elimination (exactly as in the Quadratic Sieve) to find a product of C_i's which is a perfect square.

There are several strategies that can be used to speed up CFRAC. One of them is the use of multipliers, as in the Quadratic Sieve. In fact, multipliers are sometimes essential for if the period of the continued fraction is too short then there may not be enough distinct C_i's. The large prime variation that works for the Quadratic Sieve is also appropriate for CFRAC.

There is also an early abort strategy that has proven useful. If the factor base is large, then it can be worthwhile to pause at some point in the midst of the trial division and see how far you have succeeded in reducing C_i into its factors. If very little progress has been made, then you will probably do better to abandon this C_i and look to the next.

Discussion of these strategies can be found in the papers of Morrison and Brillhart and of Pomerance listed at the end of this chapter.

11.2 Some Observations on the Bháscara-Brouncker Algorithm

The continued fraction algorithm also comes up in primality testing. To see why this is so, I want to look at part of the solution of Exercise 10.10, the first 27 lines of Algorithm 10.5 with $n = 3$.

i	A_i	B_i	C_i	P_i	Q_i
1	1	1	2	1	1
2	1	1	1	2	1
3	2	1	2	5	3
4	1	1	1	7	4
5	2	1	2	19	11
6	1	1	1	26	15
7	2	1	2	71	41
8	1	1	1	97	56
9	2	1	2	265	153
10	1	1	1	362	209
11	.	.	.	989	571
12	.	.	.	1351	780
13	.	.	.	3691	2131
14				5042	2911
15				13775	7953
16				18817	10864
17				51409	29681
18				70226	40545
19				191861	110771
20				262087	151316
21				716035	413403
22				978122	564719
23				2672279	1542841
24				3650401	2107560
25				9973081	5757961
26				13623482	7865521
27				37220045	21489003

At first glance there does not appear to be much structure to the P_i's and Q_i's. But then looking down the list of Q_i's, we may notice that Q_i is divisible by 3 if and only if i is divisible by 3, and Q_i is divisible by 4 if and only if i is divisible by 4. Q_5 is not divisible by 5, but it is by 11 and Q_i turns out to be divisible by 11 every time i is divisible by 5.

The P_i's do not seem to work quite as nicely. Two divides P_2 but it does not divide P_i whenever i is even, only when i is an odd multiple of 2.

Similarly, 5 divides P_3. It does not divide P_i whenever i is a multiple of 3, but only when i is an odd multiple of 3. These observations are summed up in the next two theorems.

Theorem 11.1 *Let P_i, Q_i be the sequences generated by Algorithm 10.5 when $n = 3$. If j is a multiple of i then Q_j is a multiple of Q_i. If j is an odd multiple of i then P_j is a multiple of P_i.*

Theorem 11.2 *Let Q_i be as in Theorem 11.1. Let m be an integer larger than 1 and let e be the smallest positive integer such that m divides Q_e. Then m divides Q_i if and only if e divides i.*

Notice how similar Theorem 11.2 sounds to Theorem 9.1 which can be stated as

> *Let e be the smallest positive integer such that m divides $b^e - 1$, then m divides $b^i - 1$ if and only if e divides i.*

While we still have not proved anything, let us assume for the moment that Theorems 11.1 and 11.2 hold. We will pursue the analogy with orders. If p is an odd prime larger than 3, then the order of 3 modulo p is a divisor of $p - 1$. Let us call e, the smallest integer such that p divides Q_e, the *rank of p*. We observe some values of e:

$$
\begin{array}{llll}
\text{the rank of} & 3 & \text{is} & 3, \\
\text{the rank of} & 5 & \text{is} & 6, \\
\text{the rank of} & 7 & \text{is} & 8, \\
\text{the rank of} & 11 & \text{is} & 5, \\
\text{the rank of} & 13 & \text{is} & 12, \\
\text{the rank of} & 17 & \text{is} & 9, \\
\text{the rank of} & 19 & \text{is} & 10, \\
\text{the rank of} & 23 & \text{is} & 22.
\end{array}
$$

It is no longer true that the rank always divides $p - 1$. For $p = 3$, the rank divides p. Sometimes the rank divides $p + 1$. This is true if p is 5, 7, 17, or 19. Comparing this list with Corollary 7.6 suggests that the rank in fact divides $p - (3/p)$ where $(3/p)$ is the Legendre symbol.

Theorem 11.3 *Let p be an odd prime, then the rank of p divides $p - (3/p)$, where $(3/p)$ is the Legendre symbol.*

Before proving these theorems, we will need two lemmas.

Lemma 11.4 *Let P_m and Q_m be as defined in Algorithm 10.5 with $n = 3$, then*

$$2^{\lfloor m/2 \rfloor} \times (P_m + Q_m \times \sqrt{3}) = (1 + \sqrt{3})^m.$$

Proof: If we compare our table of values for C_i, P_i, Q_i with Theorem 10.2, we see that

$$x_i = P_{2i} \text{ and}$$
$$y_i = Q_{2i}.$$

Making this substitution into Equation (10.3) gives us

$$P_{2i} + Q_{2i} \times \sqrt{3} = (2 + \sqrt{3})^i.$$

By using the identity

$$2 \times (2 + \sqrt{3}) = (1 + \sqrt{3})^2,$$

we get that

$$2^i \times (P_{2i} + Q_{2i} \times \sqrt{3}) = (1 + \sqrt{3})^{2i},$$

which is the statement of the theorem when m is even.

To prove the lemma for odd m, we first observe that by Theorem 10.2,

$$P_{2i+2} = 2P_{2i} + 3Q_{2i}, \tag{11.3}$$

$$Q_{2i+2} = P_{2i} + 2Q_{2i}. \tag{11.4}$$

It follows from Algorithm 10.5 that we also have

$$P_{2i+2} = P_{2i} + P_{2i+1}, \tag{11.5}$$

$$Q_{2i+2} = Q_{2i} + Q_{2i+1}. \tag{11.6}$$

Subtracting the first pair of equations from the second yields

$$P_{2i+1} = P_{2i} + 3Q_{2i}, \tag{11.7}$$

$$Q_{2i+1} = Q_{2i} + P_{2i}. \tag{11.8}$$

We now use this pair of identities and the fact that we have already proved our lemma for m even.

$$2^i \times (1 + \sqrt{3})^{2i+1} \; = \; 2^i \times (1 + \sqrt{3})^{2i} \times (1 + \sqrt{3})$$
$$= \; (P_{2i} + Q_{2i} \times \sqrt{3}) \times (1 + \sqrt{3})$$
$$= \; P_{2i} + 3Q_{2i} + (P_{2i} + Q_{2i}) \times \sqrt{3}$$
$$= \; P_{2i+1} + Q_{2i+1} \times \sqrt{3}.$$

Q.E.D.

Lemma 11.5 *Let*

$$t_i \; = \; 2^{\lfloor i/2 \rfloor} \times P_i,$$
$$u_i \; = \; 2^{\lfloor i/2 \rfloor} \times Q_i,$$

then

$$t_{i+j} = t_i \times t_j + 3u_i \times u_j, \tag{11.9}$$

$$u_{i+j} = u_i \times t_j + u_j \times t_i, \tag{11.10}$$

and if i is at least as large as j then

$$(-2)^j \times u_{i-j} = u_i \times t_j - u_j \times t_i, \tag{11.11}$$

$$(-2)^j \times t_{i-j} = t_i \times t_j - 3u_i \times u_j. \tag{11.12}$$

Proof: By Lemma 11.4 we have that

$$t_i + u_i \times \sqrt{3} = (1 + \sqrt{3})^i.$$

Therefore,

$$t_{i+j} + u_{i+j} \times \sqrt{3} = (t_i + u_i \times \sqrt{3}) \times (t_j + u_j \times \sqrt{3}).$$

Multiplying out the right-hand side and comparing the constants and co-efficients of $\sqrt{3}$ yields Equations (11.9) and (11.10).

For the remaining equations we have that

$$(t_i - u_i \times \sqrt{3}) \times (t_j + u_j \times \sqrt{3})$$
$$= (1 - \sqrt{3})^i \times (1 + \sqrt{3})^j$$
$$= (1 - \sqrt{3})^{i-j} \times (-2)^j$$
$$= (-2)^j \times (t_{i-j} - u_{i-j} \times \sqrt{3}).$$

Equations (11.11) and (11.12) follow by comparing the constant terms and coefficients of $\sqrt{3}$ as before.

Q.E.D.

11.3 Proofs of the Observations

Proof of Theorem 11.1: We will first show that if j is a multiple of i, then u_j is a multiple of u_i, and if j is an odd multiple of i then t_j is a multiple of t_i. Since t_i and u_i only differ from P_i and Q_i, respectively, by a factor which is a power of 2, it will only remain to verify that the powers of 2 are correct.

From Equation (11.10) with $i = j$ we have that

$$u_{2i} = 2u_i \times t_i, \tag{11.13}$$

therefore u_i divides u_{2i}. We proceed by induction. Assume that u_i divides u_{ri}. Again using Equation (11.10) we have that

$$u_{(r+1)i} = u_{ri} \times t_i + u_i \times t_{ri}.$$

Since u_i divides both products on the right-hand side, it also divides $u_{(r+1)i}$.

Again by Equation (11.10) with $i = j$, we see that t_i divides u_{2i}, and so by what we have just proven, t_i divides u_{2ri} for any positive integer r. By Equation (11.9) we have that

$$t_{(2r+1)i} = t_{2ri} \times t_i + 3u_{2ri} \times u_i.$$

Since t_i divides both products on the right-hand side, it also divides $t_{(2r+1)i}$.

Using Equations (11.3) and (11.4) we obtain

$$P_{i+4} = 7P_i + 12Q_i, \tag{11.14}$$
$$Q_{i+4} = 7Q_i + 4P_i \quad \text{when } i \text{ is even.} \tag{11.15}$$

Equations (11.5) and (11.6) imply that

$$P_i = P_{i+1} - P_{i-1},$$
$$Q_i = Q_{i+1} - Q_{i-1} \quad \text{when } i \text{ is odd,}$$

and so Equations (11.14) and (11.15) hold for any value of i. From these equalities and the first four values of P_i and Q_i, it follows that P_i is odd unless i is congruent to 2 modulo 4, in which case P_i is congruent to 2 modulo 4. Q_i is only even when 4 divides i. From Equation (11.10) we know that

$$Q_{4i} = 2 \times P_{2i} \times Q_{2i}. \tag{11.16}$$

If i is odd then 2 divides P_{2i} once and it does not divide Q_{2i}, so that there are exactly two factors of 2 in Q_{4i}. If i is even, say

$$i = 2^t \times \text{(odd integer)},$$

then it follows by induction using Equation (11.16) that there are exactly $2 + t$ factors of 2 in Q_{4i}.

Q.E.D.

Proof of Theorem 11.2: Let e be the rank of m and let us assume that m also divides Q_i. We need to show that i is a multiple of e. By what we have just shown in the proof of Theorem 11.1, it is enough to prove this theorem when m is odd. We can write i as

$$i = q \times e + r, \quad 0 \le r < e.$$

If $r = 0$ then i is a multiple of e. If not, then it follows from Equation (11.11) that

$$(-2)^r \times u_{i-r} = u_i \times t_r - u_r \times t_i.$$

Since m divides Q_e, it also divides $Q_{qe} = Q_{i-r}$, and thus m divides u_{i-r}. By our assumption, m divides u_i. It follows that m divides $u_r \times t_i$. Since r is less than the rank, the greatest common divisor of m and t_i must be larger than 1.

We now use Equation (11.12) with $i = j$ to get

$$(-2)^i = t_i^2 - 3u_i^2.$$

Now m is relatively prime to the left-hand side of this equation, but not to the right-hand side. This is our contradiction which implies that r must be zero.

Q.E.D.

Proof of Theorem 11.3: Let p be an odd prime. Using Lemma 11.4 and the binomial expansion, we have that

$$
\begin{aligned}
t_p + u_p\sqrt{3} &= (1+\sqrt{3})^p \\
&= 1 + p\sqrt{3} + \frac{p \times (p-1)}{1 \times 2} \times 3 \\
&\quad + \frac{p \times (p-1) \times (p-2)}{1 \times 2 \times 3} \times 3^{3/2} + \cdots \\
&\quad + p \times 3^{(p-1)/2} + 3^{p/2} \\
&\equiv 1 + 3^{(p-1)/2}\sqrt{3} \pmod{p}.
\end{aligned}
$$

Comparing constant terms and coefficients of $\sqrt{3}$, we see that

$$t_p \equiv 1 \pmod{p}, \tag{11.17}$$

$$u_p \equiv 3^{(p-1)/2} \equiv (3/p) \pmod{p}. \tag{11.18}$$

The second congruence follows from Corollary 7.1.

Using Equations (11.10) and (11.11) and the fact that $t_1 = u_1 = 1$, we get that

$$
\begin{aligned}
u_{p+1} &= u_p \times t_1 + u_1 \times t_p \\
&\equiv (3/p) + 1 \pmod{p}, \\
-2 \times u_{p-1} &= u_p \times t_1 - u_1 \times t_p \\
&\equiv (3/p) - 1 \pmod{p}.
\end{aligned}
$$

Thus if $(3/p) = 1$, then p divides u_{p-1} and so p divides Q_{p-1}. If $(3/p) = -1$, then p divides u_{p+1} and so p divides Q_{p+1}. By inspection we see that if $p = 3$, then p divides Q_p.

Q.E.D.

11.4   Primality Testing with Continued Fractions

We are now on the verge of explaining the primality test for Mersenne primes given in Algorithm 2.9. Recall from Chapter 9 that if we can find an element with order $n-1$ modulo n, then n must be prime. An analogous statement works for ranks. We first need to find the analog of the Euler function.

Definition: Let n be an odd integer with factorization given by

$$n = p_1^{a_1} \times p_2^{a_2} \times \cdots \times p_r^{a_r}.$$

We define the function $\psi(n)$ to be

$$\psi(n) = 2^{1-r} \times (p_1 - (3/p_1)) \times p_1^{a_1-1} \times \cdots \times (p_r - (3/p_r)) \times p_r^{a_r-1}.$$

Note that this looks just like the ϕ function except that we have a non-positive power of 2 out front and if 3 is not a quadratic residue modulo p_i, then we have a factor of $p_i + 1$ instead of $p_i - 1$. Just as the order of 3 mod n always divides $\phi(n)$, we shall now show that the rank of n always divides $\psi(n)$.

Lemma 11.6 *If n is a power of an odd prime, then the rank of n divides $\psi(n)$.*

Proof. By Theorem 11.2, this lemma is equivalent to saying that n divides $Q_{\psi(n)}$. Let $n = p^i$ where p is an odd prime. When $i = 1$, this lemma reduces to Theorem 11.3. We proceed by induction. We need to show that if p^i divides Q_m then p^{i+1} divides Q_{pm}.

If p^i divides Q_m, then it also divides u_m. By Lemma 11.4 and the binomial expansion, we have that

$$
\begin{aligned}
t_{pm} + u_{pm} \times \sqrt{3} &= (t_m + u_m \times \sqrt{3})^p \\
&= t_m^p + p \times t_m^{p-1} \times u_m \times \sqrt{3} \\
&\quad + \frac{p \times (p-1)}{1 \times 2} \times t_m^{p-2} \times u_m^2 \times 3 + \cdots \\
&\quad + u_m^p \times 3^{(p-1)/2} \times \sqrt{3}.
\end{aligned}
$$

Comparing coefficients of $\sqrt{3}$ on both sides yields

$$
\begin{aligned}
u_{pm} &= p \times t_m^{p-1} \times u_m + \frac{p \times (p-1) \times (p-2)}{1 \times 2 \times 3} \times t_m^{p-3} \times u_m^3 \times 3 \\
&\quad + \cdots + u_m^p \times 3^{(p-1)/2}.
\end{aligned}
$$

Since p^i divides u_m and i is at least 1, p^{i+1} divides each term on the right side of the equality above, and so it divides u_{pm}. Since p is odd and u and Q only differ by a factor which is a power of 2, p^{i+1} also divides Q_{pm}.

Q.E.D.

Lemma 11.7 *Let m and n be positive relatively prime odd integers which are not divisible by 3. Let $i = \psi(m)$ and $j = \psi(n)$, then $m \times n$ divides $Q_{ij/2}$.*

Proof: Since 3 does not divide m or n, i and j are both even, and the lemma follows from Theorem 11.1.

Q.E.D.

Theorem 11.8 *Let n be an odd integer which is not divisible by 3. Then the rank of n is a divisor of $\psi(n)$.*

Proof: This follows by induction on r, the number of distinct primes dividing n, using Lemmas 11.6 and 11.7.

Q.E.D.

Theorem 11.9 *Let n be a positive odd integer not divisible by 3 and let $(3/n)$ be the Jacobi symbol ($= 1$ if $n \equiv 1$ or -1 (mod 12), $= -1$ if $n \equiv 5$ or -5 (mod 12)). If the rank of n is $n - (3/n)$, then n is prime.*

Proof: Assume that n is composite. We first consider the case where n is a power of a prime, say $n = p^i$, i at least two. If the rank of n is $n \pm 1$ then it is relatively prime to p. On the other hand, by Theorem 11.8, the rank must divide

$$\psi(n) = (p - (3/p)) \times p^{i-1},$$

and so the rank must divide $p - (3/p)$. But the rank of n is at least $p^i - 1$ which is strictly larger than $p - (3/p)$.

If n is divisible by at least two distinct primes, say

$$n = p_1^{a_1} \times \cdots \times p_r^{a_r}, \quad \text{then}$$

$$
\begin{aligned}
\psi(n) &= 2^{1-r} \times (p_1 - (3/p_1)) \times p_1^{a_1-1} \times \cdots \times (p_r - (3/p_r)) \times p_r^{a_r-1} \\
&= 2 \times n \times \left(\frac{1}{2} - \frac{(3/p_1)}{2p_1} \right) \times \cdots \times \left(\frac{1}{2} - \frac{(3/p_r)}{2p_r} \right) \\
&\leq 2 \times n \times \left(\frac{1}{2} + \frac{1}{2p_1} \right) \times \cdots \times \left(\frac{1}{2} + \frac{1}{2p_r} \right) \\
&= 2 \times n \times \left(\frac{p_1 + 1}{2p_1} \right) \times \cdots \times \left(\frac{p_r + 1}{2p_r} \right) \\
&\leq 2 \times n \times \left(\frac{5+1}{2 \times 5} \right) \times \left(\frac{7+1}{2 \times 7} \right) \\
&= \frac{24}{35} n \\
&< n - (3/n).
\end{aligned}
$$

Thus if n is composite then $n - (3/n)$ cannot divide $\psi(n)$, and so the rank cannot be $n - (3/n)$.

Q.E.D.

Theorem 11.9 gives a primality test which is remarkably like that established in Chapter 9. If you know all of the distinct primes which divide $n - (3/n)$, then you need only verify that for each prime divisor p of $n - (3/n)$, n divides $Q_{n-(3/n)}$ and it does not divide $Q_{(n-(3/n))/p}$. Using Equations (11.9) and (11.10), we can express P_{2i} and Q_{2i} in terms of P_i and Q_i, so that there is an exact analog of our exponentiation algorithm which enables us to compute any specific Q_j in time proportional to $\log j$. Keep in mind, however, that if n is prime there is no guarantee that the rank of n will be $n - (3/n)$. All we know for certain is that the rank will divide $n - (3/n)$. Much of the time, this particular test will be inconclusive.

If 3 is a quadratic residue mod n, then this test of primality rests on being able to factor $n - 1$, and we gain nothing over the primality test established in Chapter 9. On the other hand, if 3 is not a quadratic residue mod n, then we are looking for the factorization of $n + 1$. Sometimes it is easier to factor $n + 1$ than $n - 1$. One such instance is when we want to decide if a Mersenne number $M(p) = 2^p - 1$ is prime. We will show that for all Mersenne numbers $M(p)$, the Jacobi symbol $(3/M(p))$ is -1. And as we will see in the next section, when $M(p)$ is prime its rank is $M(p) + 1$. In this particular instance, this test is always conclusive.

11.5   The Lucas-Lehmer Algorithm Explained

Theorem 11.10 *Let $M(n) = 2^n - 1$ where n is odd and at least 3. $M(n)$ is prime if and only if $M(n)$ divides $P_{(M(n)+1)/2} = P_{2^{n-1}}$.*

Proof. Since 4 is congruent to 1 mod 3, every even power of 2 is congruent to 1 mod 3, and so every odd power of 2 is congruent to 2 mod 3. This implies that $M(n) = 2^n - 1$ is congruent to 1 mod 3.

Since n is at least 3, $M(n)$ is congruent to 7 modulo 8. From these two congruences, we know that $M(n)$ is congruent to 7 modulo 24. This implies that the Jacobi symbol $(3/M(n))$ equals -1. It also implies that the Jacobi symbol $(-2/M(n))$ equals -1.

We know that P_i and Q_i are relatively prime and by Equation (11.13) that $u_{2i} = 2 \times t_i \times u_i$. Thus if $M(n)$ divides $P_{2^{n-1}}$, it must divide Q_{2^n} and it cannot divide $Q_{2^{n-1}}$. Therefore the rank of $M(n)$ is $2^n = M(n) + 1$, and by Theorem 11.9 $M(n)$ must be prime.

To prove the other direction, we first set $j = i$ in Equations (11.9) and (11.11):

$$
\begin{aligned}
t_{2i} &= t_i^2 + 3u_i^2, \\
(-2)^i &= t_i^2 - 3u_i^2.
\end{aligned}
$$

By adding this pair of equations we get

$$t_{2i} + (-2)^i = 2t_i^2. \tag{11.19}$$

Now if $M(n)$ is prime, then let $i = (M(n) + 1)/2 = 2^{n-1}$ in Equation (11.19)

$$2t_{2^{n-1}}^2 = t_{2^n} + (-2)^{2^{n-1}}.$$

Using Equation (11.9) to rewrite t_{2^n} yields

$$2t_{2^{n-1}}^2 = t_{M(n)} + 3 \times u_{M(n)} - 2 \times (-2)^{(M(n)-1)/2}.$$

From the congruences (11.17) and (11.18) as well as Corollary 7.1, this becomes

$$2t_{2^{n-1}}^2 \equiv 1 + 3 \times (3/M(n)) - 2 \times (-2/M(n)) \ (\mathrm{mod}\ M(n)).$$

Using the values for the Jacobi symbols which were established at the beginning of this proof, we see that the right-hand side is zero, and so $M(n)$ divides the left-hand side, which means it must divide $P_{2^{n-1}}$.

Q.E.D.

Equation (11.19) can be used to very speedily calculate $P_{2^{n-1}}$. It is equivalent to the following equality

$$P_{2i} = 2P_i^2 - 1, \qquad\qquad (11.20)$$

provided i is even. Let $S_t = 2P_{2^t}$, then $S_1 = 2P_2 = 4$ and the equation just given translates as

$$S_{t+1} = S_t^2 - 2. \qquad\qquad (11.21)$$

Theorem (11.10) says that $M(n)$ is a prime if and only if $M(n)$ divides S_{n-1}, and this is precisely Algorithm 2.9.

The reader has probably guessed by now that there is nothing particularly magical about the continued fraction expansion of the square root of 3. Similar properties hold for the continued fraction expansion of the square root of other positive integers which are not perfect squares. As we will see, however, there are related sequences which are easier to handle than our sequences P_i and Q_i. These are called Lucas sequences, which will be discussed in the next chapter.

REFERENCES

D. H. Lehmer, "An extended theory of Lucas' functions," *Ann. of Math.*, **31**(1930), 419-448.

D. H. Lehmer and R. E. Powers, "On factoring large numbers," *Bull. Amer. Math. Soc.*, **37**(1931), 770-776.

Michael A. Morrison and John Brillhart, "A Method of Factoring and the Factorization of F_7," *Math. of Comput.*, **29**(1975), 183-205.

Carl Pomerance, "Analysis and comparison of some integer factoring algorithms," pp. 89-139 in *Computational Methods in Number Theory, Part I*, H. W. Lenstra, Jr. and R. Tijdeman, eds., Mathematical Centre Tracts # 154, Matematisch Centrum, Amsterdam, 1982.

11.6   EXERCISES

1.11 Prove that in Algorithm 10.5, an equivalent formula for generating the next value of C_i is given by

$$C_i = \frac{n - B_i^2}{C_{i-1}}.$$

While this has the disadvantage of involving a division, that division must be exact. Especially in hand calculations, the exactness of the division can serve as a check that no errors have been made.

11.2 Write a program to implement CFRAC and use it to factor

$$31\ 61907\ 57417\ 40159.$$

11.3 What are the possible ranks of 29, 31, 37, and 41? Determine the rank of each of these four primes by just using the table of values of Q_i provided.

11.4 Compute $\psi(n)$ for $n = 35,\ 49,$ and 715.

11.5 Find the rank of 35, 49, and 715.

11.6 Define the *rank* mod 5 of m to be the smallest integer e such that m divides the Q_i generated by Algorithm 10.5 with $n = 5$. Find the rank mod 5 of every prime less than 50. Verify that it divides $p - (5/p)$.

IN ALL OF THE REMAINING EXERCISES, P_i AND Q_i ARE THE SEQUENCES GENERATED BY ALGORITHM 10.5 WITH $n = 5$.

11.7 Find a and b such that

$$P_m + Q_m\sqrt{5} = (a + b\sqrt{5})^m.$$

11.8 Find formulas like Equations (11.9) and (11.10) for P_{i+j} and Q_{i+j}.

11.9 Prove that

$$
\begin{aligned}
P_{i-j} &= (-1)^j \times (P_i \times P_j - 5 \times Q_i \times Q_j), \\
Q_{i-j} &= (-1)^j \times (Q_i \times P_j - P_i \times Q_j).
\end{aligned}
$$

11.10 Use the results of Exercises 11.8 and 11.9 to prove that if j is a multiple of i, then Q_j is a multiple of Q_i.

11.11 Use the results of Exercises 11.8 and 11.9 to prove that if j is an odd multiple of i then P_j is a multiple of P_i.

11.12 Let e be the rank mod 5 of m. Prove that m divides Q_i if and only if e divides i.

11.13 Prove that if p is any odd prime then the rank mod 5 of p divides $p - (5/p)$.

11.14 If n is odd and relatively prime to k, define the function

$$\psi(k,n) = 2n \times \Pi \left(\frac{1}{2} - \frac{(k/p)}{2p} \right),$$

where the product is over all primes p which divide n. Verify that $\psi(n) = \psi(3,n)$.

11.15 Prove that if n is a power of an odd prime then the rank mod 5 of n divides $\psi(5,n)$.

11.16 Let m and n be positive, relatively prime odd integers which are not divisible by 5. Prove that if $i = \psi(m)$ and $j = \psi(n)$ then $m \times n$ divides $Q_{ij/2}$.

11.17 Prove that if n is odd and not divisible by 5 then the rank mod 5 of n divides $\psi(5,n)$.

11.18 Prove that if n is a positive odd integer not divisible by 5 whose rank mod 5 equals $n - (5/n)$, then n is prime.

11.19 Use the results of Exercises 11.8 and 11.9 to find an algorithm that will compute P_j in time proportional to $\log j$.

11.20 Let $m = 2^n - 1$. Prove that the Jacobi symbol $(5/m)$ is -1 if and only if $n \equiv 2$ or $3 \pmod 4$.

11.21 Find an algorithm based on the results of Exercises 11.8 and 11.18 that will decide if $M(p) = 2^p - 1$ is prime when $p \equiv 3 \pmod 4$.

12

Lucas Sequences

"The Mathematicians are a sort of Frenchmen: when
you talk to them, they immediately translate it into
their own language, and right away it is something
utterly different."
– Johann Wolfgang Von Goethe

12.1 Basic Definitions

What really made everything tick in Chapter 11 was Lemma 11.4:

$$2^{\lfloor m/2 \rfloor} \times (P_m + Q_m \times \sqrt{3}) = (1 + \sqrt{3})^m.$$

Unfortunately, few continued fraction expansions satisfy such a nice relationship. It was Lucas' idea to concentrate on those sequences that do whether or not they arise from a continued fraction expansion.

Definition: Let D be an integer congruent to 0 or 1 modulo 4 which is not a perfect square and let P be an integer with the same parity as D so that 4 divides $P^2 - D$. Then the *Lucas sequences* $\{U_i\}$ *and* $\{V_i\}$ *for P and D* are defined by

$$V_i + U_i \times \sqrt{D} = 2^{1-i} \times (P + \sqrt{D})^i.$$

Note that if we take $D = 12$ and $P = 2$:

$$V_i + U_i \times 2 \times \sqrt{3} = 2 \times (1 + \sqrt{3})^i,$$

then V_i is twice t_i as defined in Chapter 11 and U_i is the same as u_i as defined in Chapter 11.

Lemma 12.1 *The Lucas sequences for P and D satisfy*

$$
\begin{aligned}
V_{i+1} &= \frac{1}{2} P \times V_i + \frac{1}{2} D \times U_i, \\
U_{i+1} &= \frac{1}{2} V_i + \frac{1}{2} P \times U_i.
\end{aligned}
$$

Proof:

$$
\begin{aligned}
V_{i+1} + U_{i+1} \times \sqrt{D} &= 2^{-i} \times (P + \sqrt{D})^{i+1} \\
&= 2^{-1} \times (V_i + U_i \times \sqrt{D}) \times (P + \sqrt{D}) \\
&= \frac{1}{2}(P \times V_i + D \times U_i + (V_i + P \times U_i) \times \sqrt{D}).
\end{aligned}
$$

The lemma now follows by comparing coefficients.

Q.E.D.

Definition: Given parameters P and D for Lucas sequences, we define the parameter Q to be

$$
Q = (P^2 - D)/4.
$$

Lemma 12.2 *For any pair of Lucas sequences, we have that*

$$
V_i^2 - D \times U_i^2 = 4Q^i.
$$

Proof:

$$
\begin{aligned}
V_i^2 - D \times U_i^2 &= (V_i + U_i\sqrt{D}) \times (V_i - U_i\sqrt{D}) \\
&= 2^{1-i} \times (P + \sqrt{D})^i \times 2^{1-i} \times (P - \sqrt{D})^i \\
&= 4 \times 4^{-i} \times (P^2 - D)^i = 4Q^i.
\end{aligned}
$$

Q.E.D.

Corollary 12.3 *Any common divisor of U_i and V_i must divide $4Q^i$.*

Theorem 12.4 *The Lucas sequences for P and D are recursively generated by*

$$
\begin{aligned}
V_0 &= 2, & U_0 &= 0, \\
V_1 &= P, & U_1 &= 1,
\end{aligned}
$$

$$
\begin{aligned}
U_{i+1} &= P \times U_i - Q \times U_{i-1}, \\
V_{i+1} &= P \times V_i - Q \times V_{i-1}.
\end{aligned}
$$

Note that if $P = 1$, $D = 5$, and $Q = -1$, then the Lucas sequence U_i is the familiar Fibonacci sequence: 1, 1, 2, 3, 5, 8, 13, ... Often we will begin by specifying the values of P and Q and thus the recursion which we want. The value of D is then given by

$$D = P^2 - 4Q.$$

Proof: The initial conditions are readily verified from the definition. Using our definition of V_i and U_i we have that

$$
\begin{aligned}
V_{i+1} + U_{i+1}\sqrt{D} &= 2^{1-(i+1)} \times (P + \sqrt{D})^{i+1} \\
&= 2^{1-i} \times (P + \sqrt{D})^{i-1} \times 2^{-1} \times (P^2 + 2P\sqrt{D} + D) \\
&= 2^{1-i} \times (P + \sqrt{D})^{i-1} \times (P^2 + P\sqrt{D} + (D - P^2)/2) \\
&= 2^{1-i} \times (P + \sqrt{D})^{i-1} \times (P^2 + P\sqrt{D} - 2Q) \\
&= 2^{1-i} \times (P + \sqrt{D})^{i} \times P - 2^{1-(i-1)} \times (P + \sqrt{D})^{i-1} \times Q \\
&= P \times (V_i + U_i\sqrt{D}) - Q \times (V_{i-1} + U_{i-1}\sqrt{D}) \\
&= (P \times V_i - Q \times V_{i-1}) + (P \times U_i - Q \times U_{i-1})\sqrt{D}.
\end{aligned}
$$

Q.E.D.

Lemma 12.5 *The following equalities hold for the Lucas sequences defined by the parameters P and D.*

$$2U_{i+j} = U_i \times V_j + U_j \times V_i, \tag{12.1}$$

$$2V_{i+j} = V_i \times V_j + D \times U_i \times U_j, \tag{12.2}$$

$$U_{2i} = U_i \times V_i, \tag{12.3}$$

$$2Q^j \times V_{i-j} = V_i \times V_j - D \times U_i \times U_j, \tag{12.4}$$

$$2Q^j \times U_{i-j} = U_i \times V_j - U_j \times V_i, \tag{12.5}$$

$$V_{2i} = V_i^2 - 2Q^i, \tag{12.6}$$

$$V_{2i+1} = V_i \times V_{i+1} - P \times Q^i. \tag{12.7}$$

Proof: From the definition of V_i and U_i we have that

$$
\begin{aligned}
V_{i+j} + U_{i+j} \times \sqrt{D} &= 2^{1-i-j} \times (P + \sqrt{D})^{i+j} \\
&= \tfrac{1}{2}(V_i + U_i \times \sqrt{D}) \times (V_j + U_j \times \sqrt{D}).
\end{aligned}
$$

Equations (12.1) and (12.2) follow by multiplying out the right-hand side and then comparing coefficients. Equation (12.3) is the special case of Equation (12.1) where $i = j$.

It follows from Lemma 12.2 that the inverse of $V_j + U_j\sqrt{D}$ is

$$
(V_j + U_j\sqrt{D})^{-1} = \frac{V_j - U_j\sqrt{D}}{4Q^j}.
$$

We therefore have that

$$
\begin{aligned}
V_{i-j} + U_{i-j} \times \sqrt{D} &= 2^{1-i+j} \times (P + \sqrt{D})^{i-j} \\
&= (V_i + U_i\sqrt{D}) \times 2 \times (V_j + U_j\sqrt{D})^{-1} \\
&= \frac{(V_i + U_i\sqrt{D}) \times (V_j - U_j\sqrt{D})}{2Q^j}.
\end{aligned}
$$

Equations (12.4) and (12.5) now follow by multiplying out the right-hand side and comparing coefficients.

If we add Equations (12.2) and (12.4) and then divide through by 2 we get

$$
V_{i+j} = V_i \times V_j - Q^j \times V_{i-j}.
$$

Equation (12.6) is the case $j = i$ while Equation (12.7) is the case where i is replaced by $i + 1$ and then j is set equal to i.

Q.E.D.

12.2 Divisibility Properties

All of the structure that we found in Chapter 11 also exists for these sequences.

Theorem 12.6 *Let $\{V_i\}$, $\{U_i\}$ be a pair of Lucas sequences and let p be an odd prime. If p divides U_i and j is any multiple of i, then p divides U_j. If p divides V_i and j is any odd multiple of i, then p divides V_j.*

Proof: The proof exactly follows that of Theorem 11.1. For the first part we use Equation (12.1). For the second part we combine Equations (12.2) and (12.3) to get that

$$2V_{j+2i} = V_j \times V_{2i} + D \times U_j \times U_i \times V_i.$$

Thus if p divides V_i and V_j, it also divides V_{j+2i}.

Q.E.D.

Theorem 12.7 *Let m be an integer which is relatively prime to $2Q$ and let e be the smallest positive integer such that m divides U_e. Then m divides U_i if and only if e divides i.*

Proof: Let m divide U_i where $i = q \times e + r$ where r is at least 0 and less than e. If r is not zero, then by Equation (12.5)

$$2Q^{qe} \times U_r = U_i \times V_{qe} - U_{eq} \times V_i.$$

Since m is relatively prime to $2Q$, it must divide U_r, contradicting the minimality of e.

Q.E.D.

Definition: Let m be an integer which is relatively prime to $2Q$. The *rank of m* (relative to P and D) is the smallest positive integer e such that m divides U_e.

Theorem 12.8 *If p is a prime which does not divide $2Q$, then the rank of p divides $p - (D/p)$ where (D/p) is the Legendre symbol. Furthermore, if p does not divide $2Q \times D$ then*

$$V_{p-(D/p)} \equiv 2 \times Q^{(1-(D/p))/2} \pmod{p}.$$

Proof: Using the definition of the Lucas sequences, we have that

$$V_p + U_p \times \sqrt{D} = 2^{1-p} \times (P + \sqrt{D})^p.$$

Since 2 has an inverse modulo p, 2^{1-p} is congruent to 1 modulo p. If we expand the right-hand side by the binomial theorem and then compare coefficients, we get that

$$
\begin{aligned}
V_p &\equiv P^p \pmod{p} \\
&\equiv P \pmod{p},
\end{aligned}
\tag{12.8}
$$

$$U_p \equiv D^{(p-1)/2} \pmod{p} \tag{12.9}$$
$$\equiv (D/p) \pmod{p}.$$

Now using the equations of Lemma 12.5, we see that

$$2U_{p+1} = U_p \times V_1 + U_1 \times V_p \tag{12.10}$$
$$\equiv (D/p) \times P + P \pmod{p}$$
$$\equiv P \times ((D/p) + 1) \pmod{p}.$$

$$2Q \times U_{p-1} = U_p \times P - V_p \tag{12.11}$$
$$\equiv (D/p) \times P - P \pmod{p}$$
$$\equiv P \times ((D/P) - 1) \pmod{p}.$$

$$2V_{p+1} = V_p \times P + D \times U_p \tag{12.12}$$
$$\equiv P^2 + D \times (D/P) \pmod{p}.$$

$$2Q \times V_{p-1} = V_p \times P - D \times U_p \tag{12.13}$$
$$\equiv P^2 - D \times (D/p) \pmod{p}.$$

If p divides D, then $(D/p) = 0$ and by Equation (12.9) p divides U_p. If $(D/p) = -1$, then p divides U_{p+1} and

$$2V_{p+1} \equiv 4Q \pmod{p}.$$

If $(D/p) = 1$, then p divides U_{p-1} and

$$2Q \times V_{p-1} \equiv 4Q \pmod{p}.$$

Q.E.D.

Henceforth, we shall assume that p does not divide $2Q$. If p does not divide D, then $p - (D/p)$ is even. Equation (12.3) then says that

$$U_{p-(D/p)} = U_{(p-(D/P))/2} \times V_{(p-(D/P))/2}.$$

By Theorem 12.8 p divides the left-hand side of this equality. By Corollary 12.3 p can divide at most one of the factors on the right-hand side. The following theorem says which one.

Theorem 12.9 *If p does not divide $2Q \times D$, then p divides $U_{(p-(D/p))/2}$ if and only if $(Q/p) = 1$.*

Proof: From Equation (12.6) we have that

$$V_i^2 = V_{2i} + 2Q^i.$$

Setting $i = (p - (D/p))/2$ and using the results from Theorem 12.8 yields

$$
\begin{aligned}
V_{(p-(D/p))/2}^2 &= V_{p-(D/p)} + 2Q^{(p-(D/p))/2} \\
&\equiv 2Q^{(1-(D/p))/2} \times (1 + Q^{(p-1)/2}) \,(\mathrm{mod}\ p) \\
&\equiv 2Q^{(1-(D/p))/2} \times (1 + (Q/p)) \,(\mathrm{mod}\ p).
\end{aligned}
$$

Thus p divides $V_{(p-(D/p))/2}$ if and only if $(Q/p) = -1$.

$$\text{Q.E.D.}$$

12.3   Lucas' Primality Test

We continue to parallel what was done in Chapter 11, now developing a primality test that relies on being able to factor $p - (D/p)$.

Definition: Given P and D, let n be a positive integer relatively prime to $2Q$ with factorization given by

$$n = p_1^{a_1} \times p_2^{a_2} \times \cdots \times p_r^{a_r}.$$

We define the function $\psi(n)$ (or $\psi(D, n)$ if the parameter D needs to be specified) to be

$$\psi(n) = 2^{1-r} \times (p_1 - (D/p_1)) \times p_1^{a_1-1} \times \cdots \times (p_r - (D/p_r)) \times p_r^{a_r-1}.$$

Lemma 12.10 *If n is a power of a prime and if n is relatively prime to $2Q$, then the rank of n divides $\psi(n)$.*

Proof: Let $n = p^i$ where p is an odd prime which does not divide Q. If $i = 1$, this lemma reduces to Theorem 12.8. We proceed by induction on i. Let us assume that p^i divides U_m. We need to be able to show that p^{i+1} divides U_{pm}.

By the definition of the Lucas sequences, we have that

$$V_{pm} + U_{pm} \times \sqrt{D} = 2^{1-p} \times (V_m + U_m \times \sqrt{D})^p.$$

If we expand the right-hand side by the binomial theorem and compare coefficients of $\sqrt{D}$, we see that

$$
2^{p-1} \times U_{pm} = p \times V_m^{p-1} \times U_m + \frac{p \times (p-1) \times (p-2)}{1 \times 2 \times 3} \times V_m^{p-3} \times U_m^3 \times D
$$
$$
+ \cdots + U_m^p \times D^{(p-1)/2}.
$$

Since p^i divides U_m and i is at least 1, p^{i+1} divides each term on the right-hand side of the above equality.

Q.E.D.

Lemma 12.11 *Let m and n be positive relatively prime integers which are each relatively prime to $2Q \times D$. Let $i = \psi(m)$ and $j = \psi(n)$, then $m \times n$ divides $U_{ij/2}$.*

Proof. Since m and n are relatively prime to D, i and j are both even. Thus $(i \times j)/2$ is a multiple of i and is also a multiple of j. It follows that both m and n divide $U_{ij/2}$, and since they are relatively prime so does their product.

Q.E.D.

Theorem 12.12 *Let n be an integer which is relatively prime to $2Q \times D$. Then the rank of n is a divisor of $\psi(n)$.*

Proof. This follows by induction on r, the number of distinct primes dividing n, using Lemmas 12.10 and 12.11.

Q.E.D.

Theorem 12.13 *Let n be a positive integer which is relatively prime to $2Q \times D$ and let (D/n) be the Jacobi symbol. If the rank of n is $n - (D/n)$ then n is prime.*

Proof. The proof is identical to that of Theorem 11.9 except that 3 is replaced by D.

Q.E.D.

In practice, what this all boils down to is the following. Suppose that n is a suspected prime and that we do not know the factorization of $n - 1$ but we do know the factorization of $n + 1$. Say,

$$n + 1 = p_1^{a_1} \times \cdots \times p_r^{a_r}.$$

We can always find a D such that $(D/n) = -1$. In practice, we choose pairs (P, Q) until we find one for which $(Q/n) = -1$ and $(D/n) = -1$ where $D = P^2 - 4Q$. We check that $\gcd(n, 2Q \times D) = 1$ and compute U_{n+1} and $U_{(n+1)/p}$ for each prime p dividing $n + 1$. If the first is divisible by n and none of the others are divisible by n, then we know that n is prime. If U_{n+1} is not divisible by n, then we know that n is composite. Otherwise, the test is inconclusive and we need to choose new values for P and Q.

This can be simplified slightly using the fact that 2 must be one of the primes dividing $n + 1$. It is enough to check that n divides $V_{(n+1)/2}$ which implies that it divides U_{n+1} but not $U_{(n+1)/2}$, and then to check for each i from 2 to r that n does not divide $V_{(n+1)/2p_i}$. Since we have chosen Q such that $(Q/n) = -1$, Theorem 12.9 tells us that if n does not divide $V_{(n+1)/2}$ then n cannot be prime. On the other hand, if n does divide $V_{(n+1)/2p_i}$ for some i between 2 and r, that only means an inconclusive choice of Lucas sequences and one must choose a different value for P or different values for both P and Q.

This test is predicated on the assumption that if n is prime, then there is some Lucas sequence for which the rank of n is $n + 1$. A proof that this assumption is valid is outlined in Exercises 13.16 and 13.17.

Brillhart, Lehmer, and Selfridge have also published several theorems similar to Pocklington's Theorem (Theorem 9.11) in which it is sufficient to partially factor $n + 1$ or to prove primality from a partial factorization of $n + 1$ together with a partial factorization of $n - 1$.

12.4 Computing the V's

Values of V_t can be computed very quickly. Equations (12.6) and (12.7) tell us how to compute V_{2i}, V_{2i+1}, and V_{2i+2} from V_i and V_{i+1}. We keep either V_{2i} and V_{2i+1} or V_{2i+1} and V_{2i+2}, depending on the binary expansion of t.

This should look very familiar. It is Algorithm 8.3 with $h = P$ and $n = Q$. We now see that we have all the ingredients for a proof of Theorem 8.2.

Proof of Theorem 8.2: The sequence v_i given in this theorem is in fact the Lucas sequence V_i for $P = h$ and $Q = n$. Since $D = P^2 - 4Q$, the condition on the Legendre symbol is simply that

$$(D/p) = -1.$$

The equation for v_{2i} is Equation (12.6). Using this equation and Theorem 12.8 with $Q = n$ we have that

$$
\begin{aligned}
V_{(p+1)/2}^2 &= V_{p+1} + 2n^{(p+1)/2} \\
&\equiv 2n + 2n \times (n/p) \pmod{p}.
\end{aligned}
$$

Now we are under the assumption that

$$
x^2 \equiv n \pmod{p}
$$

has a solution. Therefore $(n/p) = 1$, and so

$$
\begin{aligned}
V_{(p+1)/2}^2 &\equiv 4n \pmod{p}, \\
\left(\frac{p+1}{2} \times V_{(p+1)/2} \right)^2 &\equiv n \pmod{p}.
\end{aligned}
$$

Q.E.D.

One can also use the Lucas sequences to build a $p+1$ method of factorization analogous to the Pollard $p - 1$ method. This factorization method was first described by Hugh Williams.

Let us assume that n is composite and p is an unknown prime dividing n with the property that $p + 1$ is only divisible by small primes. More specifically, let us assume that $p+1$ divides 10000!. If we generate a Lucas sequence with a value of D which satisfies $(D/p) = -1$, then p will divide $U_{10000!}$ and so

$$
gcd(n, U_{10000!}) > 1.
$$

This *gcd* will be less than n as long as there is at least one prime, say q, dividing n and such that $q - (D/q)$ does not divide $U_{10000!}$.

There are two problems that we run into here that we did not encounter with the $p - 1$ method. The first involves the computation of $U_{10000!}$. We were able to compute $b^{10000!}$ very efficiently by using $b^{k!}$ MOD n to compute $b^{(k+1)!}$ MOD n in only about $\log(k+1)$ steps. In fact, we needed to compute the intermediate values in order to increase our chances of picking up one of the prime divisors of n without picking them all up.

In general, there is no simple way of computing U_{km} directly from U_m. Furthermore, Algorithm 8.3 tells us how to compute the values of V_i, not U_i. Perhaps surprisingly, we do have the following result for the V's.

Lemma 12.14 *Define* $U_i(P)$ *and* $V_i(P)$ *to be the* i^{th} *terms in the Lucas sequences with parameters* P, $Q = 1$, *and* $D = P^2 - 4$. *We then have the equalities*

$$U_{mk}(P) = U_k(P) \times U_m(V_k(P)), \qquad (12.14)$$

$$V_{mk}(P) = V_m(V_k(P)). \qquad (12.15)$$

Proof: Let $P' = V_k(P)$ and $Q' = 1$, then the corresponding D' is given by $P'^2 - 4$ and by Lemma 12.2 we can rewrite $V_k(P)^2$ in terms of D and $U_k(P)$ so that

$$D' = D \times U_k(P)^2.$$

The lemma now follows from the definition of the Lucas sequences:

$$
\begin{aligned}
V_{mk}(P) + U_{mk}(P) \times \sqrt{D} &= 2^{1-mk} \times (P + \sqrt{D})^{mk} \\
&= 2^{1-m} \times (2^{1-k} \times (P + \sqrt{D})^k)^m \\
&= 2^{1-m} \times (V_k(P) + U_k(P) \times \sqrt{D})^m \\
&= 2^{1-m} \times (P' + \sqrt{D'})^m \\
&= V_m(P') + U_m(P') \times \sqrt{D'} \\
&= V_m(P') + U_m(P') \times U_k(P) \times \sqrt{D}.
\end{aligned}
$$

Q.E.D.

We see that it is much easier to compute $V_{10000!}$ than $U_{10000!}$. Fortunately, we can use $V_{10000!}$ to locate the prime factors, p, of n for which $p+1$ divides $10000!$.

Lemma 12.15 *Let* V_i *be a Lucas sequence with parameters* P, $Q = 1$, *and* $D = P^2 - 4$. *Let* p *be a prime such that* $(D/p) = -1$ *and let* m *be a positive integer, then*

$$V_{m(p+1)} \equiv 2 \,(mod\ p).$$

Proof: The case $m = 1$ is given by Theorem 12.8. Since p divides $U_{m(p+1)}$, we have that

$$
\begin{aligned}
V_{m(p+1)} &\equiv V_{m(p+1)} + U_{m(p+1)} \times \sqrt{D} \ (\mathrm{mod}\ p) \\
&\equiv 2^{1-m} \times (2^{-p} \times (P + \sqrt{D})^{p+1})^m \ (\mathrm{mod}\ p) \\
&\equiv 2^{1-m} \times (V_{p+1} + U_{p+1} \times \sqrt{D})^m \ (\mathrm{mod}\ p) \\
&\equiv 2^{1-m} \times 2^m \equiv 2 \ (\mathrm{mod}\ p).
\end{aligned}
$$

Q.E.D.

Therefore, if $p + 1$ divides 10000! then p will divide $V_{10000!} - 2$.

We now come to our second problem. The prime p is unknown. How can we pick a P such that $D = P^2 - 4$ satisfies $(D/p) = -1$? The answer is that we cannot, not with certainty. But we do have roughly a 50-50 chance that a randomly chosen P will work, provided of course that $p + 1$ really does have only small prime divisors. In practice, one tries several different values of P, say at least three. If none of those work, then there probably is no prime divisor p of n such that $p + 1$ divides 10000!, and we move on to a different factorization technique.

Algorithm 12.16 *Williams' $p + 1$ method. Our input consists of m, the number to be factored, a randomly chosen integer P, and* **max**, *the maximum number of cycles to go through before aborting.*

```
INITIALIZE:    READ m, P, max
               count ← 1
               v ← P
```

count *is one more than the number of* v*'s that have been computed.*

```
FIND_10TH_V:   WHILE GCD(v-2,m) = 1 AND count ≤ max DO
                    FOR i = 1 to 10 DO
                            v ← NEXTV(1,P,count,m)
                            P ← v
                            count ← count + 1

TERMINATE:     WRITE GCD(v-2,m)


NEXTV(n,h,j,p):
```

This is Algorithm 8.3, returning the final value of **v** *to the caller.*

`GCD(a,b):`

This is Algorithm 1.7, returning the final value of **u** *to the caller.*

REFERENCES

John Brillhart, D. H. Lehmer and J. L. Selfridge, "New primality criteria and factorizations of $2^m \pm 1$," *Math. of Computation*, **29**(1975), 620-647.

D. H. Lehmer, "An extended theory of Lucas functions," *Annals of Math.* **31**(1930), 419-448.

Edouard Lucas, "Théorie des fonctions numériques simplement périodiques," *Amer. J. of Math.*, **1**(1878), 184-240, 289-321.

Hugh Williams, "A $p + 1$ method of factoring," *Math. of Computation*, **39**(1982), 225-234.

12.5 EXERCISES

12.1 Why is D restricted to be congruent to 0 or 1 modulo 4 in the definition of a Lucas sequence?

12.2 Prove that for any prime p,

$$(a + b)^p \equiv a^p + b^p \ (\text{mod } p).$$

12.3 Prove that U_i and V_i are always integers.

12.4 Compute the first fifty terms of the Lucas sequences for $P = 1$, $D = 5$. Compare them with the first fifty terms of P_i and Q_i in the continued fraction expansion of the square root of 5.

12.5 Compute the first fifty terms of the Lucas sequences for $P = 3$, $D = 5$. Compare them with the sequences in Exercise 12.4.

12.6 For each of the three pairs of sequences described in Exercises 12.4 and 12.5, find the rank of each of the primes $p = 5, 7, 11, 13$, and 17. Verify that in each case the rank of p divides $p - (5/p)$.

12.7 If we extend our notion of an extended integer to also include

$$\frac{a + b\sqrt{D}}{2}$$

when $D \equiv 1 \pmod 4$ and a and b are both odd, then $2 + \sqrt{5}$, $(1 + \sqrt{5})/2$, and $(3 + \sqrt{5})/2$ are all units. What is the relationship among them?

12.8 Let a and b be the roots of

$$x^2 - Px + Q = 0.$$

Prove that the Lucas sequences V_i, U_i for P and $D = P^2 - 4Q$ satisfy

$$V_i = a^i + b^i, \quad U_i = (a^i - b^i)/(a - b).$$

12.9 The Fibonacci sequence is defined by $F_1 = F_2 = 1$, $F_{i+1} = F_i + F_{i-1}$. Show that

$$F_i = \frac{(1 + \sqrt{5})^i - (1 - \sqrt{5})^i}{2\sqrt{5}}.$$

12.10 Consider the Lucas sequences for $P = 1$, $D = 5$. Use Equations (12.10)-(12.13) to compute the congruence class modulo p of U_{p-1}, U_{p+1}, V_{p-1} and V_{p+1} for each of the following primes:

$$170\,809, \quad 43054\,72081, \quad 3\,83725\,33757.$$

IN EXERCISES (12.11)-(12.14), V_i IS A LUCAS SEQUENCE FOR P AND $D = P^2 - 4$.

12.11 Prove that if p is an odd prime that does not divide D, then $V_{p-(D/p)}$ is congruent to 2 modulo p.

12.12 We shall call n a Lucas pseudoprime for the base P if n is not prime but it is relatively prime to $2D$ and $V_{n-(D/n)}$ is congruent to 2 modulo n. Find the Lucas pseudoprimes for the base 3 which are less than 500.

12.13 Prove that if p is a prime and V_{2i} is congruent to 2 modulo p, then V_i is congruent to either 2 or -2 modulo p. Show that this implies that if

$$p - (D/p) = 2^t \times s,$$

where t is at least 1 and s is odd, then p divides exactly one of the following:

$$V_s - 2, \quad V_s + 2, \quad V_{2s} + 2, \quad V_{4s} + 2, \quad \cdots, \quad V_{2^{t-1} \times s} + 2.$$

12.14 We shall call n a strong Lucas pseudoprime for the base P if it is not prime but it is relatively prime to $2D$ and it divides exactly one of the $t + 1$ expressions in Exercise 12.13. Show that a strong Lucas pseudoprime is always a Lucas pseudoprime. Find the strong Lucas pseudoprimes for the base 3 which are less than 500.

12.15 Compute $\psi(5, 38\,913)$ and $\psi(7, 738\,261)$.

12.16 Compute $\psi(D, n)$ for $D = 3$, 5, and 7 and all values of n less than 50 for which the function is defined.

12.17 Prove that if V_i is a Lucas sequence for P and $D = P^2 - 4$ and if n is relatively prime to $2D$, then for any integer m

$$V_{m \times \psi(n)+1} \equiv V_1 \pmod{n}.$$

This identity establishes an RSA-type crypto-system based on Lucas sequences. If p and q are odd primes and $n = p \times q$, choose an e which is relatively prime to $p - 1$, $p + 1$, $q - 1$, and $q + 1$. The values of n and e are published. The encoder converts his message to a number P less than n. The encoded message is then the term V_e of the Lucas sequence for P and $D = P^2 - 4$, where we can safely assume that n and D are relatively prime. If d is the inverse of e modulo $\psi(n)$, then $V_{de} \equiv V_1 \equiv P \pmod{n}$. Furthermore, by Lemma 12.14, V_{de} is the dth term in the Lucas sequence with $P = V_e$ and $D = V_e^2 - 4$, so that it is easily computed.

The only drawback to this version of the RSA crypto-system is that the value of $\psi(n)$ depends on the values of (D/p) and (D/q), which depend on the message being sent. However, there are only two possible values for each, so only four possible values of $\psi(n)$. One computes the four inverses of e, one for each possible value of $\psi(n)$. One of those four values of d will successfully decode the incoming message.

12.18 For the Lucas sequence with $P = 5$, $Q = 1$, compute V_i for each of the following three values of i:

$$1067, \quad 235\,061, \quad 892\,466\,712.$$

12.19 Use the Lucas sequence primality test to prove that each of the following is really a prime:

$$8779, \quad 98\,641, \quad 290\,249.$$

12.20 Use the Lucas sequence primality test to prove that

$$14357\,40214\,80139$$

is prime.

12.21 Use Williams' $p + 1$ algorithm to factor:

$$41\,953\,267 \quad \text{and} \quad 24839\,76259.$$

12.22 Use Williams' $p + 1$ algorithm to factor:

$$69\,70561\,65709.$$

13

Groups and Elliptic Curves

"The Theory of Groups is a branch of mathematics in which one does something to something and then compares the result with the result obtained from doing the same thing to something else, or something else to the same thing."
– James R. Newman (The World of Mathematics)

13.1 Groups

Something has been going on in the past few chapters that we should get such very similar factorization techniques and primality tests from constructions as distinct as exponentiation and Lucas sequences. If we can understand the general framework of what works, it may help us find more specific examples in which the tests are faster and more efficient. With that hope in mind, we introduce some notation.

Definition: A *group*, G is a set together with a binary operation, say ∂, such that

1. The operation is *closed*. If x and y are in G, then $x\partial y$ is also in G.

2. The operation is *associative*. If x, y, and z are in G, then $(x\partial y)\partial z = x\partial(y\partial z)$.

3. G contains an *identity*, say e. For each x in G, $x\partial e = e\partial x = x$.

4. Each element of G has an *inverse*. If x is in G, then there is a y in G such that $x\partial y = y\partial x = e$.

The integers with addition form a group. Zero is the identity and $-x$ is the inverse of x. This group is called $\mathbf{Z}$.

If n is any positive integer, then the positive integers less than and relatively prime to n together with multiplication modulo n form a group. For example, if $n = 12$ then 1 is the identity, 5, 7, and 11 are the remaining elements and they are each their own inverses. If $n = 9$ then 1 is the identity, 5 is the inverse of 2, 7 is the inverse of 4, and 8 is its own inverse. Lemma 4.1 guarantees that we always have a unique inverse. We shall call this group $U(\mathbf{Z}/n\mathbf{Z})$.

For a much more complicated example, we consider the group that sits behind what is happening in the Lucas sequences when $Q = 1$, what we will call $L(D, n)$ where D is congruent to 0 or 1 modulo 4 and n is relatively prime to $2D$. The elements of $L(D, n)$ are pairs of residues modulo n, say (a, b), satisfying

$$a^2 - D \times b^2 \equiv 4 \,(\text{mod } n).$$

If (a, b) and (x, y) are both elements of $L(D, n)$, then the binary operation is defined by setting

$$\alpha = \frac{n+1}{2} \times (a \times x + D \times b \times y) \text{ MOD } n,$$

$$\beta = \frac{n+1}{2} \times (b \times x + a \times y) \text{ MOD } n,$$

and then

$$(a, b)\partial(x, y) = (\alpha, \beta).$$

It is left as an exercise to verify that this operation is closed, associative, that $(2,0)$ is the identity, and that the inverse of (a, b) is $(a, -b)$.

Definition: The *order of a group* G, denoted by $|G|$, is the number of elements in G.

For the three examples given above, the respective orders are $|\mathbf{Z}| = \infty$, $|U(\mathbf{Z}/n\mathbf{Z})| = \phi(n)$, and $|L(D, n)| = 2^{r-1} \times \psi(D, n)$ where r is the number of distinct primes dividing n and $\psi(D, n)$ is the function defined in Section 12.3. The order of $L(D, n)$ is derived in the exercises.

Definition: Given a group G with binary operation ∂, identity e, and an element x in G, we define the *powers of x in G* as follows:

$$
\begin{aligned}
x\#-1 &= \text{the inverse of } x, \\
x\#0 &= e, \\
x\#1 &= x, \\
x\#2 &= x\partial x, \\
x\#3 &= x\partial x\partial x,
\end{aligned}
$$

and in general

$$x\#i = x\partial(x\#(i-1)) = (x\#-1)\partial(x\#(i+1)).$$

Definition: The *order of an element* x in G is the smallest positive integer i such that

$$e = x \# i.$$

Lemma 13.1 *If the group G has finite order, then every x in G has a finite order.*

Proof: Since $x \# i$ is always in G and G has only finitely many elements, we can find two positive integers $i < j$ such that

$$x \# i = x \# j = (x \# i) \partial (x \# (j - i)).$$

Since $x \# i$ is in G, it has an inverse and so

$$e = x \# (j - i).$$

Q.E.D.

Theorem 13.2 *If x has order i, then $x \# j = e$ if and only if i divides j.*

Proof: If $j = m \times i$, then

$$x \# j = (x \# i) \# m = e \# m = e.$$

If $x \# j = e$, then write $j = m \times i + r$ where $0 \le r < i$. Then

$$e = x \# j = (x \# (m \times i)) \partial (x \# r) = x \# r.$$

Therefore $r = 0$ by the minimality of i.

Q.E.D.

Theorem 13.3 *If x is an element of G then the order of x divides the order of G.*

Proof: Let H_1 be the set of all elements of G which can be written in the form $x \# i$ where i is a positive integer. The number of elements in H_1 is the order of x. Note that x, the inverse of x, and e are all in H_1. If H_1 is all of G then the orders of x and G are equal and the theorem holds.

If H_1 is not all of G, let a be an element of G which is not in H_1. Let H_2 be the set of elements of the form $a \partial (x \# i)$ where i is a positive integer less than or equal to the order of x. If there are positive integers i and j less than or equal to the order of x and such that

$$a\partial(x\#i) = a\partial(x\#j),$$

then

$$x\#i = x\#j,$$

which implies that i equals j. Thus the elements of H_2 are distinct. If any element of H_2 were also in H_1, say

$$a\partial(x\#i) = x\#j,$$

then

$$a = x\#(j - i),$$

and so a would be in H_1. Therefore all elements of H_1 and H_2 are distinct and the number of elements in the union of H_1 and H_2 is twice the order of x. If there are no other elements in G, then the order of G is twice the order of x and the theorem holds.

We proceed by induction. Assume that we have constructed k sets $H_1, H_2, \ldots, H_k$ such that each set consists of elements of the form $b\partial(x\#i)$ where b is an element of G which does not appear in any of the preceding sets and i is a positive integer less than or equal to the order of x. Furthermore, we assume that these k sets have distinct elements so that their union has $k \times$ (order of x) elements. Finally, we assume that there is an element of G, say c, which is not in this union. We form the set H_{k+1} of elements of the form $c\partial(x\#i)$ where i is a positive integer less than or equal to the order of x. The elements of H_{k+1} are distinct and if any of them were in a previous set, then c would be in that set. Therefore, the union of our $k + 1$ sets contains $(k + 1) \times$ (order of x) distinct elements. If this exhausts G, then the theorem holds. If not, then we can repeat our inductive step. Since the order of G is finite, we must eventually exhaust all of G.

Q.E.D.

13.2   A General Approach to Primality Tests

Let n be a candidate for primality and assume that we have a group G whose elements are a subset of the residues modulo n or some subset of vectors of residues modulo n. Let us further assume that the possible orders of elements of G depend on the factorization of n in such a way that an element of order m can exist if and only if n is prime. We have this situation in $U(\mathbf{Z}/n\mathbf{Z})$ where an element can only have order $n-1$ if n is prime. We also have it in $L(D, n)$ where the order of each element divides $\psi(D, n)$ and so if we can find an element of order $n+1$ or $n-1$ then n must be prime.

If we know the factorization of m, we can prove that an element x in G has order m if we can verify that

$$x\#m = e \quad \text{and} \quad x\#(m/p) \neq e$$

for every prime p dividing m.

This procedure can be modified in the manner of Pocklington's theorem (Theorem 9.11). We first define two terms.

Definition: We say that G *is a group modulo* n if its elements are vectors of residues modulo n and its binary operation is defined in terms of arithmetic operations modulo n. If d is any divisor of n, then the *restricted group modulo* d, denoted $G|d$, is the group derived from G by reducing each coordinate modulo d.

If n is composite then it is divisible by a prime less than or equal to the square root of n. Call this prime q. In both of our examples, G is a group modulo n and the order of $G|q$ is at most $q+1$ which is at most one more than the square root of n. If m is at least two more than the square root of n, then every element of $G|q$ has order strictly less than m.

If there is an element x in G such that

$$x\#m = e,$$

and for every prime p dividing m some coordinate of $x\#(m/p) - e$ is relatively prime to n, then the order of x in $G|q$ is m, contradicting the assumption that all elements of $G|q$ have order strictly less than m. Thus n must be prime.

In practice, we may have to try several different elements before we find one which has order m. In our first example, this means choosing different bases b. In the second example, it means trying different choices of P. In the general situation, we want the group G to have a lot of elements of order m so that our odds of hitting one by chance are fairly high. Note that we do not want to invoke a primality test of this form unless we have a very high confidence that our number really is prime. If x does not have order m, it might be because n is composite or it might be only due to a poor choice of x.

We sum up this approach in the following theorem.

Theorem 13.4 *Let n be a suspected prime and assume that we have a group modulo n, say G. Let $G|d$ be the restricted group modulo d and let e*

be the identity in G. If we can find an element x in G and an integer m satisfying the following conditions, then n is prime:

(1) The integer m is larger than the order of $G|q$ would be for any prime q dividing n and less than the square root of n.

(2) $x\#m = e$.

(3) For each prime p dividing m, some coordinate of $x\#(m/p) - e$ is relatively prime to n.

13.3   A General Approach to Factorization

Both the Pollard $p-1$ method and the Williams $p+1$ method fit into this context of groups whose elements are vectors of residues modulo n. We now let n be a number which is known to be composite and p be an unknown prime divisor of n. Let G be a group modulo n and $G|p$ the restricted group modulo p.

If the order of $G|p$ is considerably less than the order of G, then we can hope to find an element x in G and an integer k such that $x\#(k!)$ is not the identity in G, but the corresponding computation in $G|p$ does yield the identity of $G|p$. This means that there is at least one coordinate of $x\#(k!) - e$ which is not divisible by n, but all of the coordinates are divisible by p. Taking the greatest common divisor of n and the coordinate which is not divisible by n will yield a non-trivial divisor of n.

In our first example, we choose a base b and compute

$$gcd(b^{k!} - 1, n).$$

In the second example, we choose the parameter P, set $Q = 1$, and for the resulting Lucas sequence compute

$$gcd(V_{k!} - 2, n).$$

Note that since we do not know p in advance, we also do not know what value of k will work. In practice, we compute the *gcd* at regular intervals, say for every tenth value of k.

Like the primality test, this algorithm could go on for a long time without yielding an answer. If n really is prime, it will only come up with inconclusive results. There is also no way of knowing *a priori* that the order of $G|p$ will divide $k!$ for some prime p dividing n. We observe in addition that this approach requires an efficient means of calculating at least one coordinate of $x\#(k!)$.

We sum up this approach in the following theorem.

Theorem 13.5 *Let n be a composite number and let G be a group modulo n. Let p be a prime dividing n and let $G|p$ be the restricted group modulo p. If the order of $G|p$ divides $k!$, then p divides each coordinate of $x\#(k!)-e$. If n does not divide the tth coordinate of $x\#(k!)-e$, then the greatest common divisor of n and the tth coordinate of $x\#(k!)-e$ is a non-trivial divisor of n.*

13.4 Elliptic Curves

All of this theory would be wasted if we did not have any more examples of groups modulo n with all the desired properties. But there is at least one more, it arises out of the arithmetic of elliptic curves.

Consider the equation

$$y^2 = x^3 + ax + b, \tag{13.1}$$

where a and b are constants chosen so that

$$4a^3 + 27b \neq 0. \tag{13.2}$$

This merely guarantees that the cubic equation

$$z = x^3 + ax + b, \tag{13.3}$$

has three distinct roots. Equation (13.1) is only solvable for y when the right-hand side is positive, and then y is merely $+$ or $-$ the square root of the right-hand side. If Equation (13.3) has three real roots, then the graph of Equation (13.1) looks like Figure 2.

This curve has the curious property that if a non-vertical line intersects it at two points, then it will also have a third point of intersection. A tangent to the curve is considered to have two points of intersection at the point of tangency. We can compute the extra point of intersection using the following lemma.

Lemma 13.6 *Let (x_1, y_1) and (x_2, y_2) be two points on the elliptic curve given by*

$$y^2 = x^3 + ax + b, \quad 4a^3 + 27b^2 \neq 0.$$

We assume that if $x_1 = x_2$ then $y_1 \neq -y_2$. We do, however, permit the two points to be the same provided $y_1 \neq 0$. The third point of intersection, (x_3, y_3), is calculated in the following manner:
If $x_1 \neq x_2$, then set

$$\lambda = \frac{y_1 - y_2}{x_1 - x_2}.$$

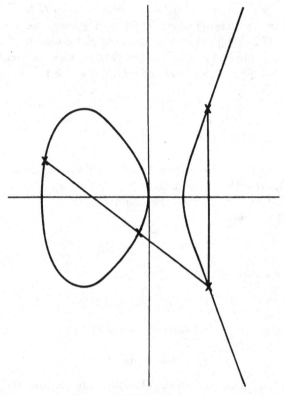

FIGURE 13.1.

If $x_1 = x_2$, then set

$$\lambda = \frac{3x_1^2 + a}{2y_1}.$$

We then have that

$$\begin{aligned}
x_3 &= \lambda^2 - x_1 - x_2, \\
y_3 &= \lambda \times (x_3 - x_1) + y_1.
\end{aligned}$$

Proof. The quantity λ is the slope of the line connecting our two points. This is clear if $x_1 \neq x_2$. It needs to be proven if the x's are equal. Since both points satisfy Equation (13.1), we have that

$$\begin{aligned}
y_1^2 - y_2^2 &= x_1^3 - x_2^3 + a \times (x_1 - x_2), \\
(y_1 - y_2) \times (y_1 + y_2) &= (x_1 - x_2) \times (x_1^2 + x_1 \times x_2 + x_2^2 + a),
\end{aligned}$$

$$\lambda = \frac{y_1 - y_2}{x_1 - x_2} = \frac{x_1^2 + x_1 \times x_2 + x_2^2 + a}{y_1 + y_2}. \tag{13.4}$$

As x_2 approaches x_1, the right-hand side approaches

$$\frac{3x_1^2 + a}{2y_1}.$$

Equation (13.4) holds for any pair of points on our line, so we also have that

$$\begin{aligned}
\lambda \times (y_3 + y_1) &= x_3^2 + x_3 \times x_1 + x_1^2 + a, \\
\lambda \times (y_3 + y_2) &= x_3^2 + x_3 \times x_2 + x_2^2 + a.
\end{aligned}$$

Subtracting the second equation from the first yields

$$\lambda \times (y_1 - y_2) = x_3 \times (x_1 - x_2) + (x_1^2 - x_2^2).$$

Dividing both sides by $x_1 - x_2$ gives us that

$$\lambda \times \lambda = x_3 + x_1 + x_2,$$

from which we can calculate x_3. The calculation of y_3 follows from the definition of the slope of our line.

Q.E.D.

It should be clear from Lemma 13.6 that if both points have rational coordinates then so does the third point.

Definition: Given an elliptic curve and two rational points on that curve: (x_1, y_1) and $(x_2, y_2) \neq (x_1, -y_1)$, we define a binary operation by

$$(x_1, y_1)\partial(x_2, y_2) = (x_3, -y_3),$$

where x_3 and y_3 are defined by Lemma 13.6. Note that the sum of two points is *not* the third point on that line, but the reflection across the x-axis of that third point as shown in Figure 2. It is still on the same elliptic curve.

We now have a set, namely the rational points on an elliptic curve, and a binary operation. We would like to make this into a group. To do this, we need to define $(x, y)\partial(x, -y)$, we need to find an identity, and we have to find inverses. We can solve all our problems with one stroke.

Definition: We define ∞ to be the identity for the binary operation ∂ and define

$$(x, y)\partial(x, -y) = (x, -y)\partial(x, y) = \infty.$$

The point ∞ can be thought of as a point infinitely far north so that every vertical line passes through it. One of the beauties of this definition is that now *every* straight line which intersects the curve at two points also intersects at a third.

Definition: Given an elliptic curve,

$$y^2 = x^3 + ax + b, \quad 4a^2 + 27b^2 \neq 0,$$

let $E(a, b)$ denote the group of rational points on the curve together with the point ∞ at infinity with the binary operation ∂ as defined above.

13.5   Elliptic Curves Modulo p

Definition: All of our arithmetic operations make perfectly good sense modulo n, provided that the denominators are relatively prime to n. Specifically, we define the operation ∂ *modulo* n by

$$\infty \quad \text{is the identity.}$$

If $x_1 \equiv x_2 \pmod{n}$ and $y_1 \equiv -y_2 \pmod{n}$, then

$$(x_1, y_1)\partial(x_2, y_2) = \infty.$$

If $x_1 \not\equiv x_2 \pmod{n}$ and if $gcd(x_1 - x_2, n) = 1$, then let s be the inverse of $x_1 - x_2$ modulo n and define λ by

$$\lambda = (y_1 - y_2) \times s \text{ MOD } n.$$

If $x_1 \equiv x_2 \pmod{n}$ and if $gcd(y_1 + y_2, n) = 1$, then $y_1 \equiv y_2 \pmod{n}$, so let s be the inverse of $2y_1$ modulo n and define λ by

$$\lambda = (3 \times x_1^2 + a) \times s \text{ MOD } n.$$

Define x_3 and y_3 by

$$\begin{aligned} x_3 &= (\lambda^2 - x_1 - x_2) \text{ MOD } n, \\ y_3 &= (\lambda \times (x_3 - x_1) + y_1) \text{ MOD } n. \end{aligned}$$

The binary operation ∂ modulo n is then given by

$$(x_1, y_1)\partial(x_2, y_2) \equiv (x_3, -y_3) \pmod{n},$$

when x_3 and y_3 are defined. In particular, if n is an odd prime then our binary operation is always defined.

Definition: Let p be a prime larger than 3 and let a and b be integers chosen such that

$$4a^3 + 27b^2 \not\equiv 0 \pmod{p}.$$

Then $E(a, b)/p$ denotes the elliptic group modulo p whose elements are pairs (x, y) of non-negative integers less than p satisfying

$$y^2 \equiv x^3 + ax + b \pmod{p},$$

together with the identity, ∞, and whose binary operation is given by ∂ modulo p as defined above.

The machinery developed in Section 13.2 and 13.3 can now be brought into play. Given (x_1, y_1) in $E(a, b)/p$, we define

$$(x_i, y_i) \equiv (x_1, y_1) \# i \pmod{p}.$$

As an example, let $p = 5$, $a = b = -1$. The points of $E(-1, -1)/5$ must satisfy

$$y^2 \equiv x^3 - x - 1 \pmod{5}.$$

Note that x cannot be 3 because 23 is not a quadratic residue modulo 5. The elements of our group are

$(0, 2)$, $(1, 2)$, $(2, 0)$, $(4, 2)$,

$(0, 3)$, $(1, 3)$, $(4, 3)$,   and ∞.

If $(x_1, y_1) = (0, 2)$, then

$$
\begin{aligned}
(x_2, y_2) &= (0,2)\partial(0,2) & \lambda &\equiv (3 \times 0 - 1) \times 4 &\equiv 1 \pmod 5, \\
&& x_2 &\equiv 1 - 0 - 0 &\equiv 1 \pmod 5, \\
&& -y_2 &\equiv 1 \times (1 - 0) + 2 &\equiv 3 \pmod 5, \\
&= (1,2);
\end{aligned}
$$

$$
\begin{aligned}
(x_3, y_3) &= (1,2)\partial(0,2) & \lambda &\equiv (2 - 2) \times 1 &\equiv 0 \pmod 5, \\
&& x_3 &\equiv 0 - 1 - 0 &\equiv 4 \pmod 5, \\
&& -y_3 &\equiv 0 \times (4 - 0) + 2 &\equiv 2 \pmod 5, \\
&= (4,3);
\end{aligned}
$$

$$
\begin{aligned}
(x_4, y_4) &= (4.3)\partial(0,2) & \lambda &\equiv (3 - 2) \times 4 &\equiv \pmod 5, \\
&& x_4 &\equiv 16 - 4 - 0 &\equiv 2 \pmod 5, \\
&& -y_4 &\equiv 4 \times (2 - 0) + 2 &\equiv 0 \pmod 5, \\
&= (2,0);
\end{aligned}
$$

$$
\begin{aligned}
(x_5, y_5) &= (2,0)\partial(0,2) & \lambda &\equiv (0 - 2) \times 3 &\equiv 4 \pmod 5, \\
&& x_5 &\equiv 16 - 2 - 0 &\equiv 4 \pmod 5, \\
&& -y_5 &\equiv 4 \times (4 - 0) + 2 &\equiv 3 \pmod 5, \\
&= (4,2);
\end{aligned}
$$

$$
\begin{aligned}
(x_6, y_6) &= (4,2)\partial(0,2) & \lambda &\equiv (2 - 2) \times 4 &\equiv 0 \pmod 5, \\
&& x_6 &\equiv 0 - 4 - 0 &\equiv 1 \pmod 5, \\
&& -y_6 &\equiv 0 \times (1 - 0) + 2 &\equiv 2 \pmod 5, \\
&= (1,3);
\end{aligned}
$$

$$
\begin{aligned}
(x_7, y_7) &= (1,3)\partial(0,2) & \lambda &\equiv (3 - 2) \times 1 &\equiv 1 \pmod 5, \\
&& x_7 &\equiv 1 - 1 - 0 &\equiv 0 \pmod 5, \\
&& -y_7 &\equiv 1 \times (0 - 0) + 2 &\equiv 2 \pmod 5, \\
&= (0,3);
\end{aligned}
$$

$$
(x_8, y_8) = (0,3)\partial(0,2) = \infty.
$$

Just as in our previous techniques for factorization and primality testing, the key to elliptic curve methods lies in knowing the order of $E(a,b)/p$. This order can be evaluated by observing that for every residue class modulo p, if $x^3 + ax + b$ is a quadratic residue, then there are two values of y that correspond to that x, if $x^3 + ax + b$ is divisible by p, then there is one value of y that corresponds to that x, and otherwise there are no values of y that correspond to that x. Since we also have one point at infinity, we can express the order of $E(a,b)/p$ in terms of the Legendre symbol:

$$
|E(a,b)/p| = 1 + \sum_{x=1}^{p} \left(\left(\frac{x^3 + ax + b}{p} \right) + 1 \right).
$$

Unfortunately, this formula is totally impractical for large values of p. Nevertheless, we do know a lot about the possible order of $E(a,b)/p$, some of which we state in the next two theorems whose proofs are well beyond the scope of this book. The first of these is due to Helmut Hasse (1898-1979) and was published in 1934; the second was first proved by William Waterhouse in 1969.

Theorem 13.7 *The order of $E(a,b)/p$ lies in the interval*

$$I(p) = (p + 1 - 2\sqrt{p}, p + 1 + 2\sqrt{p}).$$

Theorem 13.8 *Given a prime p larger than 3 and any integer n in the interval $I(p)$, there exists a and b such that*

$$|E(a,b)/p| = n.$$

Furthermore, the orders of the groups of elliptic curves modulo p are fairly uniformly distributed over the interval $I(p)$.

One other aspect of the groups $U(\mathbf{Z}/n\mathbf{Z})$ and $L(D,n)$ that enabled our primality tests was the existence of elements of orders near the order of the group. Results exist which promise that this also holds for $E(a,b)/p$.

We are in a very nice situation here. Our previous groups modulo p had order $p+1$ or $p-1$. In order to prove primality we had to be able to factor $p+1$ or $p-1$. In order to find a prime divisor q of n, we needed to have either $q-1$ or $q+1$ divide $K!$ for suitable K. Our group orders now lie in a much larger range. To prove primality, we need only find an integer in the interval $I(p)$ which we can factor. To find a prime divisor q of n we only need to have some integer in the interval $I(q)$ divide $K!$.

However, the situation is not quite as idyllic as it sounds. It is not simple to determine the order of $E(a,b)/p$, nor given an integer n in $I(p)$ is it simple to find a and b such that $|E(a,b)/p| = n$. We shall have to rely on the fact that the orders are fairly uniformly distributed. We will randomly choose elliptic curves until we find one that works. All this will be made clearer in the next chapter.

REFERENCES

A. K. Lenstra and H. W. Lenstra, Jr., Algorithms in number thoery, University of Chicago, Department of Computer Science, Technical Report # 87-008, 1987.

Carl Pomerance, "Very short primality proofs," *Math. of Computation*, **48**(1987), 315-322.

13.6 EXERCISES

13.1 Prove that $L(D, n)$ is closed.

13.2 Prove that $L(D, n)$ is associative.

13.3 Prove that in $L(D, n)$, $(2,0)$ is the identity and $(a, -b)$ is the inverse of (a, b).

13.4 Let G be a group. An element g in G is called a *generator* if every element of G can be written as $g\#i$ for some integer i. Prove that if G has finite order, then g is a generator if and only if the order of g equals $|G|$.

13.5 Prove that if x has order j, then $x\#i$ has order

$$\frac{j}{gcd(i, j)} = \frac{lcm(i, j)}{i}.$$

13.6 Prove that if a finite group G has a generator, then it has exactly $\phi(|G|)$ generators.

FOR EXERCISES 13.7 - 13.10, LET $P(n)$ BE THE SET OF *PERMU-TATIONS* ON $J(n) = \{1, 2, \ldots, n\}$. THAT IS THE SET OF 1 TO 1 FUNCTIONS FROM $J(n)$ TO $J(n)$. WE MAKE IT INTO A GROUP BY USING COMPOSITION AS THE BINARY OPERATION:

$$s \circ t(i) = s(t(i)).$$

13.7 Let s, t be the elements of $P(4)$ given by

$$s(1) = 3, \qquad s(2) = 1, \qquad s(3) = 4, \qquad s(4) = 2,$$
$$t(1) = 1, \qquad t(2) = 3, \qquad t(3) = 4, \qquad t(4) = 2.$$

These permutations can be conveniently coded as $s = 3142$, $t = 1342$. Find $s \circ t$, $t \circ s$, $s\#2$, $s\#3$, $t\#2$, $t\#3$.

13.8 What is the identity of $P(n)$? Prove that every element of $P(n)$ has a unique inverse.

13.9 Show that $|P(2)| = 2$, $|P(3)| = 6$, $|P(4)| = 24$, and in general, $|P(n)| = n!$.

13.10 Find the order of each element in $P(3)$.

13.11 Prove that if p is a prime that does not divide $2D$ then

$$a^2 - Db^2 \equiv 4 \pmod{p} \qquad (*)$$

has $p - (D/p)$ solutions. (*Hint:* If $b \equiv 0 \pmod{p}$, then there are two possible values for a. If p does not divide b, then we can rewrite this congruence as

$$D \equiv (a \times b^{-1} - 2 \times b^{-1}) \times (a \times b^{-1} + 2 \times b^{-1}) \pmod{p}.$$

Show that there is a 1 to 1 correspondence between solutions of Equation $(*)$ where p does not divide b and pairs of integers r, s such that r is not congruent to s modulo p and

$$r \times s \equiv D \pmod{p}.)$$

13.12 Prove that if
$$a^2 - Db^2 \equiv 4 \pmod{p^n} \qquad (**)$$

has m solutions, then

$$a^2 - Db^2 \equiv 4 \pmod{p^{n+1}}$$

has $p \times m$ solutions. (*Hint:* Let x, y be a solution of Equation $(**)$. Show that there are exactly p pairs of residue classes modulo p, say (j, k), such that

$$(x + jp^n)^2 - D(y + kp^n)^2 \equiv 4 \pmod{p^{n+1}}.)$$

13.13 Show that if m and n are relatively prime, then the number of solutions of

$$a^2 - Db^2 \equiv 4 \pmod{m \times n}$$

is the product of the number of solutions modulo m times the number of solutions modulo n.

13.14 Pull together exercises 13.11 - 13.13 to prove that if $gcd(n, 2D) = 1$, then

$$|L(D, n)| = n \times \Pi \left(1 - \frac{(D/p)}{p} \right),$$

where the product is over all primes p which divide n.

13.15 Show that if n is divisible by r distinct primes, then the order of any element in $L(D, n)$ divides $\psi(D, n) = 2^{1-r} \times n \times \Pi \left(1 - \frac{(D/p)}{p} \right)$.

13.16 Show that the order of an element (a, b) in $L(D, n)$ is the smallest positive integer, say e, for which

$$\left(\frac{a + b\sqrt{D}}{2}\right)^e \equiv 1 \ (\text{mod } n).$$

13.17 Use Exercise 13.16 to show that if p is prime, then there exists at least one element of $L(D, p)$ whose order is $p - (D/P)$. Show that this implies that there are $\phi(p - (D/p))$ elements of order $p - (D/p)$.

13.18 In Theorem 13.4, prove that n will be prime if for each prime p dividing m there is an x such that $x \# m = e$ and $x \# (m/p) - e$ is relatively prime to n in at least one coordinate. (In other words, we do not have to find one x that works for all values of p.)

13.19 Graph the elliptic curves:

$$
\begin{aligned}
y^2 &= x^3 - 9x, \\
y^2 &= x^3 - 3x - 2, \\
y^2 &= x^3 - 3x + 2, \\
y^2 &= x^3 - 9x + 12.
\end{aligned}
$$

13.20 In $E(-2, 5)$, compute

$$(2, 3)\partial(1, 2), \quad (1, 2)\partial(1, 2), \quad (1, 2)\partial(-2, 1).$$

13.21 Note what happens if we ignore the condition $4a^3 + 27b^2 \neq 0$. Show that in $E(-3, -2)$, if $(x, y) \neq (-1, 0)$ then

$$(-1, 0)\partial(x, y) = (-1, 0).$$

13.22 In $E(0, 3)$, compute $(1, 2)\#i$ for i in $\{2, 3, \ldots, 10\}$.

13.23 In $E(0, 4)/5$, find the order of $(3, 1)$.

13.24 What is the order of $E(1, 2)/5$?

13.25 For each pair (a, b) such that $4a^3 + 27b^2 \not\equiv 0 \ (\text{mod } 5)$, find the order of $E(a, b)/5$.

14

Applications of Elliptic Curves

"And there he plays extravagant matches
In fitless finger-stalls
On a cloth untrue
With a twisted cue
And elliptical billiard balls."

– William S. Gilbert (The Mikado, Act II)

14.1   Computation on Elliptic Curves

If we are going to be able to implement factorization techniques and primality tests using the arithmetic of elliptic curves, we need a fast way of computing $(x, y)\#i$. Actually, the fastest techniques just compute the first coordinate.

Lemma 14.1 *In the elliptic group $E(a, b)$, let $(p, q) = (x, y)\partial(x, y)$, if $y \neq 0$ then*

$$p = \frac{(x^2 - a)^2 - 8bx}{4(x^3 + ax + b)}.$$

Proof: From Lemma 13.6 we have that

$$
\begin{aligned}
p &= \lambda^2 - 2x \\
&= \frac{(3x^2 + a)^2}{(2y)^2} - 2x \\
&= \frac{(3x^2 + a)^2 - 2x \times 4 \times (x^3 + ax + b)}{4(x^3 + ax + b)} \\
&= \frac{(x^2 - a)^2 - 8bx}{4(x^3 + ax + b)}.
\end{aligned}
$$

Q.E.D.

Thus, given the first coordinate of $(x, y)\#i$, we can compute the first coordinate of $(x, y)\#2i$. The next lemma shows us how to compute the first coordinate of $(x, y)\#(2i+1)$ from the first coordinates of $(x, y)\#i$ and $(x, y)\#(i+1)$.

Lemma 14.2 *In the elliptic group* $E(a, b)$, *let* $(p, q) = (x, y)\#i$, $(r, s) = (x, y)\#(i+1)$ *and* $(u, v) = (x, y)\#(2i+1)$, *if* $p \neq r$ *and* $x \neq 0$ *then*

$$u = \frac{(a - pr)^2 - 4b \times (p + r)}{x \times (p - r)^2}.$$

Proof: Since $(u, v) = (p, q)\partial(r, s)$, we can use Lemma 13.6 to get that

$$u = \frac{(q - s)^2}{(p - r)^2} - p - r,$$

$$
\begin{aligned}
u \times (p - r)^2 &= (q - s)^2 - (p + r) \times (p - r)^2 \qquad (14.1) \\
&= -2qs + 2b + (a + pr) \times (p + r).
\end{aligned}
$$

We also have that

$$(x, y) = (r, s)\partial(p, -q),$$

and therefore, by a similar argument,

$$x \times (p - r)^2 = 2qs + 2b + (a + pr) \times (p + r). \qquad (14.2)$$

Multiplying Equations (14.1) and (14.2) yields

$$
\begin{aligned}
&x \times u \times (p - r)^4 \\
&= (2b + (a + pr) \times (p + r))^2 - 4 \times (p^3 + ap + b) \times (r^3 + ar + b) \\
&= ((a - pr)^2 - 4 \times b \times (p + r)) \times (p - r)^2.
\end{aligned}
$$

The lemma now follows by dividing through by $x \times (p - r)^4$.

Q.E.D.

We can avoid rational numbers and restrict our attention to integers if we introduce a third coordinate and write

$$x = X/Z; \quad y = Y/Z, \qquad (14.3)$$

where X, Y, and Z are now integers satisfying

$$(Y/Z)^2 = (X/Z)^3 + a(X/Z) + b,$$

or equivalently

$$Y^2 Z = X^3 + aXZ^2 + bZ^3. \tag{14.4}$$

Observe that in view of Equation (14.3), the solutions (X, Y, Z) and (cX, cY, cZ) of Equation (14.4) represent the same rational solution (x, y) for any non-zero value of c. For this reason, we shall consider the two triples given above to be equal:

$$(X, Y, Z) = (cX, cY, cZ) \quad \text{for any non-zero integer } c.$$

This notation has an added bonus beyond just enabling us to work with integers, it gives us an explicit representation for the identity element. The identity corresponds to the new solution of Equation (14.4) where $Z = 0$ (and thus $X = 0$ and Y can be any integer).

The computational rules given in Lemmas 14.1 and 14.2 are restated in terms of X and Z in the following theorem.

Theorem 14.3 *Let (X, Y, Z) be an integral solution of Equation (14.4) and define (X_i, Y_i, Z_i) by*

$$(X_i/Z_i, Y_i/Z_i) = (X/Z, Y/Z)\#i.$$

We shall also write this relationship as

$$(X_i, Y_i, Z_i) = (X, Y, Z)\#i.$$

We then have the following computational rules:

$$
\begin{aligned}
X_{2i} &= (X_i^2 - a \times Z_i^2)^2 - 8b \times X_i \times Z_i^3, \\
Z_{2i} &= 4Z_i \times (X_i^3 + a \times X_i \times Z_i^2 + b \times Z_i^3), \\
X_{2i+1} &= Z \times [(X_i \times X_{i+1} - a \times Z_i \times Z_{i+1})^2 \\
&\quad -4b \times Z_i \times Z_{i+1} \times (X_i \times Z_{i+1} + X_{i+1} \times Z_i)], \\
Z_{2i+1} &= X \times (X_{i+1} \times Z_i - X_i \times Z_{i+1})^2.
\end{aligned}
$$

Proof: From Lemma 14.1 we have that

$$
\begin{aligned}
\frac{X_{2i}}{Z_{2i}} &= \frac{((\frac{X_i}{Z_i})^2 - a)^2 - 8b \times (\frac{X_i}{Z_i})}{4((\frac{X_i}{Z_i})^3 + a \times (\frac{X_i}{Z_i}) + b)} \\
&= \frac{(X_i^2 - a \times Z_i^2)^2 - 8b \times X_i \times Z_i^3}{4Z_i \times (X_i^3 + a \times X_i \times Z_i^2 + b \times Z_i^3)}.
\end{aligned}
$$

Since X_{2i} and Z_{2i} are determined up to multiplication by a constant, we can take the numerator of the expression on the right-hand side as X_{2i} and the denominator as Z_{2i}.

The proof of the equalities for X_{2i+1} and Z_{2i+1} follow from Lemma 14.2 in a similar fashion.

Q.E.D.

While we shall not need the value for Y_i, it is worth noting that, up to sign, Y_i can be recovered from the values of X_i and Z_i by using Equation (14.4).

The computational rules given in Theorem 14.3 are equally valid for the operation ∂ modulo n provided we do all of our computations modulo n. We put these rules into the following algorithm for computing the first and third coordinates in $(X, Y, Z)\#k$ modulo n.

Algorithm 14.4 *This algorithm uses the binary expansion of k and the computational rules of Theorem 14.3 to compute the first and third coordinates in $(X, Y, Z)\#k$ modulo n.*

```
INITIALIZE:    READ X, Z, k, n, a, b
               i ← 0
               WHILE k > 0 DO
                   i ← i + 1
                   C_i ← k MOD 2
                   k ← ⌊k/2⌋
               length ← i
               X1 ← X
               Z1 ← Z
               X2 ← X_SUB_2I(X,Z)
               Z2 ← Z_SUB_2I(X,Z)
```

We convert **k** *into its binary representation:*

$$k = C_1 + C_2 \times 2 + C_3 \times 4 + \cdots + C_{length} \times 2^{length-1}.$$

The initial values of (X1,Z1) *and* (X2,Z2) *are* (X_1, Z_1) *and* (X_2, Z_2), *respectively.*

```
COMPUTE_LOOP:    FOR i = length - 1 to 1 BY -1 DO
                     U1 ← X_SUB_2I_PLUS_1(X1,Z1,X2,Z2)
                     U2 ← Z_SUB_2I_PLUS_1(X1,Z1,X2,Z2)
                     IF C_i = 0 THEN DO
                         temp ← X_SUB_2I(X1,Z1)
                         Z1 ← Z_SUB_2I(X1,Z1)
                         X1 ← temp
                         X2 ← U1
                         Z2 ← U2
                     ELSE DO
                         temp ← X_SUB_2I(X2,Z2)
                         Z2 ← Z_SUB_2I(X2,Z2)
                         X2 ← temp
                         X1 ← U1
                         Z1 ← U2


TERMINATE:       WRITE X1, Z1
```

$X\_SUB\_2I(r,s)$:
$$term \leftarrow r^2 - a \times s^2 \ MOD \ n$$
$$value \leftarrow term^2 - 8 \times b \times r \times s^3 \ MOD \ n$$
RETURN value

Return value *to caller.*

$Z\_SUB\_2I(r,s)$:
$$term \leftarrow r^3 + a \times r \times s^2 + b \times s^3 \ MOD \ n$$
$$value \leftarrow 4 \times s \times term \ MOD \ n$$
RETURN value

Return value *to caller.*

$X\_SUB\_2I\_PLUS\_1(r,s,u,v)$:
$$term1 \leftarrow r \times u - a \times s \times v \ MOD \ n$$
$$term2 \leftarrow b \times s \times v \times (r \times v + s \times u) \ MOD \ n$$
$$value \leftarrow Z \times (term1^2 - 4 \times term2) \ MOD \ n$$
RETURN value

Return value *to caller.*

```
Z_SUB_2I_PLUS_1(r,s,u,v):
                    term ← u × s - r × v MOD n
                    value ← X × term² MOD n
                    RETURN value
```

Return value *to caller.*

14.2   Factorization with Elliptic Curves

The following procedure for factoring integers by means of elliptic curves is essentially due to A. K. Lenstra and H. W. Lenstra, Jr. Let n be a composite number relatively prime to 6. In practice, n is known to have no small prime factors. We randomly choose a parameter a for our elliptic curve and a point (x, y) on the curve,

$$0 \leq x, y < n.$$

Note that the values of a, x, and y uniquely determine b:

$$b \equiv y^2 - x^3 - ax \pmod{n}.$$

We verify that

$$gcd(4a^3 + 27b^2, n) = 1,$$

if not then we have probably found a factor of n. Converting to triples, (X, Y, Z), our initial triple is $(x, y, 1)$.

If p is a prime dividing n and $|E(a, b)/p|$ divides $k!$, then

$$(X, Y, Z)\#k! = (\dots(((X, Y, Z)\#1)\#2)\#3\dots)\#k$$

will be the identity in $E(a, b)/p$, which means that p will divide $Z_{k!}$. If k is not too large, there is an excellent chance that $gcd(Z_{k!}, n)$ is a non-trivial divisor of n.

Algorithm 14.5 *Factorization by elliptic curves. Let* n *be the integer to be factored,* n *must be relatively prime to 6. The constants* X, Y, *and* a *are arbitrary integers,* X *and* Y *are non-negative and less than* n. MAX *is the maximal value for* k *before aborting.*

```
INITIALIZE:      READ n, X, Y, a, MAX
                 b ← Y × Y - X × X × X - a × X MOD n
                 g ← GCD(4 × a × a × a + 27 × b × b, n)
                 IF g ≠ 1 THEN CALL TERMINATE
                 Z ← 1
                 k ← 2


COMPUTE_LOOP:    WHILE k ≤ MAX DO
                    FOR i = 1 to 10 DO
                       CALL NEXTVALUES(X,Z,k)
                       k ← k + 1
                    g ← GCD(Z,n)
                    IF g ≠ 1 THEN CALL TERMINATE


TERMINATE:       WRITE g
```

If g = 1, then MAX was not high enough to find the first prime divisor of n. If g = n, then we have picked up all the prime divisors of n. In either of these cases, start over with different values of X, Y, a. If g ≠ 1 or n, then n factors as n = g × n/g.

NEXTVALUES(X,Z,k)

This is Algorithm 14.4 with n, a, and b fixed. Return final values of X1 and Z1 as the new values of X and Z, respectively.

As in the Pollard $p - 1$ and Williams $p + 1$ methods, you can speed this algorithm by restricting k to a set of powers of primes less than MAX rather than running over all integers less than MAX. Also as with the other methods, you can expect better results if you regularly interrupt the run and restart with a new set of parameters rather than grinding on with your initial choice.

14.3 Primality Testing

Primality testing with elliptic curves follows the same principles we have developed for our previous primality tests. Let n be a suspected prime. If it really is prime, then $|E(a, b)/n|$ lies in the interval $(n + 1 - 2\sqrt{n}, n + 1 + 2\sqrt{n})$. Furthermore, it is known that if n is prime, then there are always many elements of high order. The statement I have just made is very imprecise, suffice it to say that by it I mean that there are enough elements of sufficiently high orders to make the primality test I am about to describe practical.

If we can factor $|E(a,b)/n|$, say

$$|E(a,b)/n| = q_1^{a_1} \times \cdots \times q_r^{a_r},$$

then we choose a point $P = (X_0, Y_0, Z_0)$ at random and find its order. The order must be a divisor of $|E(a,b)/n|$ and so we can start by checking $P\#(|E(a,b)/n|/q_i)$ for each i. If the Z coordinate is relatively prime to n for each i and if n divides the Z coordinate of $P\#|E(a,b)/n|$, then the order of P in $E(a,b)/p$ is $|E(a,b)/n| \geq n+1-2\sqrt{n}$ for any prime p dividing n. It follows that n must be prime.

There is still hope even if one or more of the Z coordinates is divisible by n. For each such i, we find the smallest positive integer b_i such that the Z coordinate of

$$P\# \left(\frac{|E(a,b)/n|}{q_i^{b_i}} \right)$$

is relatively prime to n. The order of P in $E(a,b)/p$ will then be at least

$$|E(a,b)/n| \times q_1^{1-b_1} \times q_2^{1-b_2} \times \cdots \times q_r^{1-b_r},$$

for any prime p dividing n. By Theorem 13.4, as long as this order is larger than $1 + \sqrt{n} + 2 \times \sqrt[4]{n}$, n must be prime.

The problem of primality testing with elliptic curves has come down to evaluating the order of $E(a,b)/n$. Unfortunately, this is not easy. As we saw in Section 13.5, the most direct formula for computing the order involves finding and summing n terms. The first elliptic curve primality test, proposed by S. Goldwasser and J. Kilian in 1986, used something called the division points algorithm to find $|E(a,b)/n|$. While this method is much faster, it still appears to be impractical for large values of n.

The first practical means of computing $|E(a,b)/n|$ for large n was described by A. K. Lenstra and H. W. Lenstra, Jr. in 1987 and is based on an idea suggested to them by A. O. L. Atkin. It relies very heavily on twentieth-century mathematics and in particular a notion that goes by the name of "elliptic curves with complex multiplication." To set it up in its full power necessitates evaluating certain theta functions and is way beyond the scope of this book. However, I would like to convey some of its flavor by showing how it works in some very special cases. Specifically, if n is congruent to 1 modulo 4, we shall determine the orders of four of the elliptic groups modulo n. And if n is congruent to 1 modulo 3, we shall see how to determine the orders of six elliptic groups.

Even these special cases rely on some very powerful mathematical results which will take us the next two sections to understand without even attempting to prove all of them. In Section 14.4 we shall get a glimpse of what has happened in the past 200 years in extending the results of Chapter 10. In Section 14.5, we shall briefly pick up again the thread we dropped at the end of Chapter 9.

14.4 Quadratic Forms

In Chapter 10, we looked for integer solutions of the equation

$$x^2 - dy^2 = \pm 1. \tag{14.5}$$

Fermat, Euler, and other mathematicians of the seventeenth and eighteenth centuries ran across other quadratic polynomials in x and y for which they needed integer solutions. By the end of the eighteenth century, mathematicians such as Lagrange, Legendre, and Gauss had started the systematic study of what are now called *quadratic forms*, polynomials of the form

$$ax^2 + bxy + cy^2,$$

where a, b, and c are integers.

Many questions can be asked about quadratic forms: For what integers n does

$$ax^2 + bxy + cy^2 = n \tag{14.6}$$

have a solution in integers? How many solutions are there? How are the solutions related? How can a specific solution be found?

We will be asking the last question about two particular quadratic forms:

$$x^2 + y^2 = n \quad \text{and} \quad x^2 + xy + y^2 = n,$$

where n is the number we want to test for primality. The next theorem tells us precisely when we have a solution.

Theorem 14.6 *If p is a prime larger than 3 then the equation*

$$x^2 + y^2 = p$$

has a solution in integers if and only if p is congruent to 1 modulo 4. The equation

$$x^2 + xy + y^2 = p$$

has a solution in integers if and only if p is congruent to 1 modulo 3.

It is a straightforward exercise to see that in the first case p is congruent to 1 modulo 4, and in the second p is congruent to 1 modulo 3. I shall show that solutions exist when the appropriate congruence is satisfied by showing how to construct a solution.

Certain quadratic forms are very closely related. As an example, consider

$$x^2 + 2xy + 2y^2 = (x + y)^2 + y^2.$$

If we can find a solution to

$$x^2 + 2xy + 2y^2 = 29,$$

then we can use it to find a solution to

$$x^2 + y^2 = 29,$$

and vice-versa. This motivates the following definitions.

Definition: There are three *basic transformations* of a quadratic form: replacing x by $x + ky$ where k is an integer, replacing x by $-x$, and interchanging x and y. We say that two quadratic forms are *equivalent* if it is possible to pass from one to the other by a sequence of basic transformations.

If $P(x, y)$ and $Q(x, y)$ are equivalent quadratic forms, then the number of solutions of $P(x, y) = n$ is the same as the number of solutions of $Q(x, y) = n$. Furthermore, if we know a sequence of transformations that will take us from $P(x, y)$ to $Q(x, y)$ and if we know a solution of $P(x, y) = n$, then we can use it to find a solution of $Q(x, y) = n$.

As an example, the equation

$$13x^2 + 10xy + 2y^2 = 13$$

has the solution $x = 1$, $y = 0$. This quadratic form is equivalent to $x^2 + y^2$ by the following sequence of basic transformations:

Interchange x and y :	$2x^2 + 10xy + 13y^2,$
Replace x by $x - 2y$:	$2x^2 + 2xy + y^2,$
Interchange x and y :	$x^2 + 2xy + 2y^2,$
Replace x by $x - y$:	$x^2 + y^2.$

Taking our original solution $(1, 0)$ through this sequence of transformations, it becomes

$$(1, 0) \rightarrow (0, 1) \rightarrow (2, 1) \rightarrow (1, 2) \rightarrow (3, 2).$$

One of the remarkable properties of the basic transformations is that they do not change the value of the *discriminant* of the quadratic form

$$D = b^2 - 4ac.$$

This property is left as an exercise. It should be noted that the discriminant is always congruent to 0 or 1 modulo 4.

Lemma 14.7 *If two quadratic forms are equivalent, then they have the same discriminant.*

Unfortunately, not all quadratic forms with the same discriminant are equivalent. When a is positive and D is negative we say that the quadratic form is *positive* (or sometimes *positive definite*). There is an efficient algorithm for determining whether two positive quadratic forms are equivalent and for finding a sequence of basic transformations that will take you from one to the other. The term "positive" comes from the fact that if a is positive and D is negative, then the quadratic form is always strictly positive unless $x = y = 0$.

Theorem 14.8 *Every positive quadratic form is equivalent to exactly one* **reduced form**, *that is a form whose coefficients satisfy:*

$$c \geq a \geq b \geq 0.$$

While I will not prove this theorem for you, I will show you how to find this unique reduced form. A proof of this theorem is outlined in Exercises 14.8, 14.11, and 14.12. If a is larger than c, then exchange x and y. If b does not lie between a and 0, replace x by $x + ky$ where k is chosen so that the new value of b lies between $-a$ and a. If the new value of b is negative, then replace x by $-x$. Now iterate these steps until the form is reduced. This is exactly what I was doing in the example given above.

If two positive quadratic forms are equivalent then we can always pass from one to the other by using this algorithm and passing through the unique reduced quadratic form.

The next algorithm, which we will need for the elliptic curve primality test, accomplishes this reduction and at the same times transforms a known solution of when the original quadratic form equals n to a solution of when the reduced quadratic form equals n.

Algorithm 14.9 *Given a, b, c, x, y, and n satisfying*

$$ax^2 + bxy + cy^2 = n,$$

this algorithm finds the equivalent reduced quadratic form and values of x and y at which it equals n.

```
INITIALIZE:        READ a, b, c, x, y
                   D ← b × b - 4 × a × c
                   CALL CHECK
```

CHECK *verifies that we really have a positive quadratic form.*

```
REDUCTION_LOOP:    WHILE b < 0 or a < b or c < a DO
                       IF c < a THEN DO
                          temp ← a
                          a ← c
                          c ← temp
                          temp ← x
                          x ← y
                          y ← temp
                       IF a < |b| THEN DO
                          k ← ⌊(a + b)/(2a)⌋
                          c ← b - 2 × k × a
                          c ← (b × b - D)/(4 × a)
                          x ← x + k × y
                       IF b < 0 THEN DO
                          b ← -b
                          x ← -x
```

If c is less than a, then we interchange x and y. If a is less than |b|, then we replace x by x + ky where 2ka lies between b - a and b + a. If b is negative, then we change the sign of x. This is all iterated until we have a reduced form.

```
TERMINATE:         WRITE a, b, c, x, y

CHECK:             IF a < 0 OR D ≥ 0 THEN DO
                      WRITE ''ERROR''
                      CALL TERMINATE
                   RETURN
```

In a reduced positive quadratic form, we have that

$$a \le c = \frac{b^2 - D}{4a} \le \frac{a^2 - D}{4a}$$

and so

$$4a^2 \le a^2 - D,$$

which implies that

$$a \le \sqrt{|D|/3}.$$

For the two quadratic forms in which we are interested, the discriminant is -4 or -3, respectively. This forces a to be 1 and b to be 0 or 1. The parameters a, b, and D uniquely determine c, and so there is only one reduced form for each of these two discriminants. Thus every quadratic form of discriminant -4 is equivalent to $x^2 + y^2$ and every quadratic form of discriminant -3 is equivalent to $x^2 + xy + y^2$. If we can find a quadratic form of the right discriminant for which we can solve equation (14.6), then we can use Algorithm 14.9 to transform that solution into one that we are looking for. The quadratic forms we are looking for are given in the next lemma.

Lemma 14.10 *If D is congruent to 0 or 1 modulo 4 and if p is an odd prime for which D is a quadratic residue, then find b such that*

$$b^2 \equiv D \ (mod \ p)$$

and b has the same parity as D. It then follows that $(b^2 - D)/4p$ is an integer and

$$px^2 + bxy + \left(\frac{b^2 - D}{4p}\right)y^2$$

is a quadratic form which has discriminant D and which is equal to p when $x = 1$, $y = 0$.

The proof of this lemma is just the calculation of the discriminant. What is important is that -4 is a quadratic residue precisely when p is congruent to 1 modulo 4, and -3 is a quadratic residue precisely when p is congruent to 1 modulo 3. We can find the value of b by using Algorithm 8.3. If the resulting b does not have the same parity as D, then we use $p - b$ instead.

As an example, to solve

$$x^2 + y^2 = 673,$$

we first solve

$$b^2 \equiv -4 \ (\text{mod } 673),$$

using Algorithm 8.3. The solutions are $b = 116$ or 557 modulo 673. We choose $b = 116$, then $(b^2 - D)/4p = (116^2 - -4)/4 \times 673 = 5$. This gives us the quadratic form

$$673x^2 + 116xy + 5y^2,$$

which has discriminant -4 and equals 673 when $x = 1$, $y = 0$. Running this through Algorithm 14.8 produces the solution $x = 23$, $y = 12$ to our original problem.

14.5 The Power Residue Symbol

The Legendre symbol (n/p), encodes whether or not n is a perfect square modulo p. In the nineteenth century, a number of mathematicians worked on the problem of extending this notion to higher powers. The work began with Gauss who found but did not publish the results on the third and fourth power residue symbols. Ferdinand Gotthold Eisenstein (1823-1852) published several proofs of these results and Ernst Edward Kummer (1810-1893) extended the results to arbitrary power residue symbols. We know from Corollary 9.5 that if d divides $p - 1$, then n is a perfect d^{th} power modulo p if and only if

$$n^{(p-1)/d} \equiv 1 \;(\text{mod } p).$$

We have also seen that as long as p does not divide n, we can define the Legendre symbol by $(n/p) = (-1)^i$ where i is the unique integer, modulo 2, such that

$$n^{(p-1)/2} \equiv (-1)^i \;(\text{mod } p).$$

To get the right analog for the dth power, we need to extend our integers to include the dth root of unity:

$$\zeta = e^{2\pi i/d}.$$

There are a number of complications that enter at this point. One of them is that numbers that were prime as ordinary integers will suddenly factor in our extended integers.

Let us take as an example $d = 4$. The fourth root of 1 is $i = \sqrt{-1}$ and we are looking at complex integers: $a + bi$. From Section 14.4, we know that any prime p which is congruent to 1 modulo 4 can be written as $a^2 + b^2$. But in our extended integers, this factors as

$$p = (a - bi) \times (a + bi).$$

This means that ordinary primes like 5, 13, and 17 are no longer prime in the complex integers:

$$
\begin{aligned}
5 &= (1 + 2i) \times (1 - 2i), \\
13 &= (3 + 2i) \times (3 - 2i), \\
17 &= (1 + 4i) \times (1 - 4i).
\end{aligned}
$$

The next result promises that we do not have to worry about our primes breaking up any further.

Theorem 14.11 *The primes in the system of complex integers are (up to multiplication by ± 1 or $\pm i$) the ordinary primes which are congruent to 3 modulo 4, $1+i$, and the complex integers $a \pm bi$ where $a^2 + b^2$ is an ordinary prime congruent to 1 modulo 4. If p is a prime in the system of complex integers and p does not divide 2, then among $\pm p$ and $\pm ip$, exactly one of these four primes in congruent to 1 modulo $2 + 2i$ (in the sense that $2 + 2i$ exactly divides the difference between that prime and 1).*

Remember that the Legendre symbol is only defined when the bottom parameter is prime, the same is true here. The fourth power symbol is not defined when the bottom parameter is an ordinary prime congruent to 1 modulo 4. Before defining the fourth power symbol, we need one more idea.

Definition: The *conjugate* of the complex integer $a + bi$ is

$$\overline{a + bi} = a - bi.$$

The *norm of a complex integer*, $a + bi$, is

$$N(a + bi) = (a + bi) \times \overline{(a + bi)} = a^2 + b^2.$$

Definition: If p is a prime in the system of complex integers, and if p does not divide either 2 or the complex integer n, then the *fourth power symbol*, $(n/p)_4$, is defined to be i^j where j is the unique integer modulo 4 satisfying

$$n^{(N(p)-1)/4} \equiv i^j \pmod{p},$$

or, equivalently, p divides $n^{(N(p)-1)/4} - i^j$. While j is well defined, I shall not prove that it is.

It is worth noting that if p is a complex prime which does not divide 2, then the norm of p is always congruent to 1 modulo 4.

As an example, to compute $(1 + 2i/3)_4$, we observe that

$$(1 + 2i)^{(9-1)/4} = -3 + 4i \equiv i \pmod{3},$$

so that

$$(1 + 2i/3)_4 = i.$$

To compute $(3/1 + 2i)_4$, we observe that $1 + 2i$ divides 5 and therefore

$$
\begin{aligned}
3^{(5-1)/4} = 3 &\equiv 3 - 3 - 6i \pmod{1 + 2i} \\
&\equiv -6i \equiv -i \pmod{1 + 2i},
\end{aligned}
$$

so that

$$(3/1 + 2i)_4 = -i.$$

We can now say something about the order of certain elliptic curves modulo p. Both this theorem and Theorem 14.14 were proved by André Weil in 1952.

Theorem 14.12 *Let n be an ordinary prime which is congruent to 1 modulo 4 and let p be a complex prime that divides n and is congruent to 1 modulo $2 + 2i$. If D is any integer not divisible by n then the order of $E(-D, 0)/n$ is*

$$|E(-D, 0)/n| = n + 1 - \overline{(D/p)_4} \times p - (D/p)_4 \times \bar{p}.$$

As an example, take $n = 13$ and $p = 3 + 2i$.

$$
\begin{aligned}
|E(-1, 0)/13| &= 14 - (3 + 2i) - (3 - 2i) = 8, \\
|E(1, 0)/13| &= 14 - (-1) \times (3 + 2i) + (-1) \times (3 - 2i) = 20, \\
|E(-2, 0)/13| &= 14 - (i) \times (3 + 2i) - (-i) \times (3 - 2i) = 18, \\
|E(2, 0)/13| &= 14 - (-i) \times (3 + 2i) - (i) \times (3 - 2i) = 10.
\end{aligned}
$$

For the next result, we consider the sixth power symbol, which means that we are working with extended integers of the form

$$a + b\omega + c\omega^2,$$

where $\omega = e^{2\pi i/3}$. Since $\omega^3 = 1$ and $\omega \neq 1$, we have that

$$0 = \frac{\omega^3 - 1}{\omega - 1} = 1 + \omega + \omega^2,$$

and so $\omega^2 = -1 - \omega$, which means that

$$a + b\omega + c\omega^2 = (a - c) + (b - c)\omega.$$

For this reason, we will write all integers in this extended system of *cubic integers* as $a + b\omega$ where a and b are ordinary integers.

Definition: The *conjugate* of the cubic integers $a + b\omega$ is

$$\overline{a + b\omega} = a + b\omega^2 = (a - b) - b\omega.$$

The *norm* of the cubic integer $a - b\omega$ is

$$N(a - b\omega) = (a - b\omega) \times (a - b\omega^2) = a^2 + ab + b^2.$$

From Section 14.4, we know that any ordinary prime which is congruent to 1 modulo 3 will factor in the cubic integers. For example,

$$13 = 4^2 + 4 \times (-3) + (-3)^2 = (4 + 3\omega) \times (1 - 3\omega),$$
$$19 = 2^2 + 2 \times (-5) + (-5)^2 = (2 + 5\omega) \times (-3 - 5\omega).$$

Theorem 14.13 *The primes in the system of cubic integers are (up to multiplication by ± 1, $\pm \omega$, or $\pm \omega^2$) the ordinary primes which are congruent to 2 modulo 3, $1 - \omega$, and the cubic integers $a - b\omega$ and $a - b\omega^2$ where $a^2 + ab + b^2$ is an ordinary prime congruent to 1 modulo 3. If p is a prime in the system of cubic integers and p does not divide 3, then among $\pm p$, $\pm \omega p$, and $\pm \omega^2 p$, exactly one of these six primes is congruent to 2 modulo 3.*

Definition: If p is a prime in the system of cubic integers, and if p does not divide either 6 or the cubic integer n, then the *sixth power symbol*, $(n/p)_6$, is defined to be $(-\omega)^j$ where j is the unique integer modulo 6 satisfying

$$n^{(N(p)-1)/6} = (-\omega)^j \pmod{p},$$

or, equivalently, p divides $n^{(N(p)-1)/6} - (-\omega)^j$. While j is well-defined, I shall not prove that it is.

If $n = -4 - 3\omega$ and $p = 5$, then we have

$$
\begin{aligned}
(-4 - 3\omega)^{(25-1)/6} &= 256 + 768\omega + 864\omega^2 + 432\omega^3 + 81\omega^4 \\
&= (688 - 864) + (849 - 864)\omega \\
&\equiv 4 \pmod{5}, \\
&\equiv -1 \pmod{5},
\end{aligned}
$$

so that $(-4 - 3\omega/5)_6 = -1$.

If $n = 5$ and $p = -4 - 3\omega$ then we have

$$5^{(13-1)/6} = 25 \equiv -1 \pmod{-4 - 3\omega},$$

so that $(5/-4 - 3\omega)_6 = -1$.

Theorem 14.14 *Let n be an ordinary prime which is congruent to 1 modulo 3 and let p be a cubic prime that divides n and is congruent to 2 modulo 3. If D is any integer not divisible by n then the order of $E(0, D)/n$ is*

$$|E(0, D)/n| = n + 1 + \overline{(4D/p)_6} \times p + (4D/p)_6 \times \bar{p}.$$

As an example, take $n = 13$ and $p = -4 - 3\omega$:

$$
\begin{aligned}
|E(0,1)/13| &= 14 + (\omega^2) \times (-4 - 3\omega) + (\omega) \times (-1 + 3\omega) = 12, \\
|E(0,2)/13| &= 14 + (-1) \times (-4 - 3\omega) + (-1) \times (-1 + 3\omega) = 19, \\
|E(0,3)/13| &= 14 + (1) \times (-4 - 3\omega) + (1) \times (-1 + 3\omega) = 9, \\
|E(0,4)/13| &= 14 + (\omega) \times (-4 - 3\omega) + (\omega^2 \times (-1 + 3\omega) = 21, \\
|E(0,5)/13| &= 14 + (-\omega^2) \times (-4 - 3\omega) + (-\omega) \times (-1 + 3\omega) = 16, \\
|E(0,6)/13| &= 14 + (-\omega) \times (-4 - 3\omega) + (-\omega 2) \times (-1 + 3\omega) = 7.
\end{aligned}
$$

REFERENCES

Kenneth Ireland and Michael Rosen, *A Classical Introduction to Modern Number Theory*, Springer-Verlag, New York, 1982.

A. K. Lenstra and H. W. Lenstra, Jr., Algorithms in number theory, University of Chicago, Department of Computer Science, Technical Report # 87-008, 1987.

Peter L. Montgomery, "Speeding the Pollard and Elliptic Curve Methods of Factorization," *Math. of Computation*, **48**(1987), 243-264.

14.6 EXERCISES

PETER MONTGOMERY HAS SUGGESTED USING THE FOLLOWING REPRESENTATION FOR ELLIPTIC CURVES:

$$By^2 = x^3 + Ax^2 + x, \quad B \times (A^2 - 4) \neq 0.$$

EXERCISES 14.1 - 14.4 DEVELOP THE COMPUTATIONAL RULES FOR THIS REPRESENTATION.

14.1 Let (x_1, y_1), (x_2, y_2), and (x_3, y_3) be the three points of intersection of a straight line of slope λ with an elliptic curve in the Montgomery form. Prove that

$$\lambda^2 = x_1 + x_2 + x_3 + a.$$

14.2 Let (x, y) be a point on an elliptic curve in the Montgomery form and let $(p, q) = (x, y)\#2$. Show that

$$p = \frac{(x^2 - 1)^2}{4x(x^2 + Ax + 1)}.$$

14.3 Let (x, y) be a point on an elliptic curve in the Montgomery form and let $(p, q) = (x, y)\#i$, $(r, s) = (x, y)\#(i + 1)$, $(u, v) = (x, y)\#(2i + 1)$. Show that

$$u = \frac{(pr - 1)^2}{x(p - r)^2}.$$

14.4 Rewrite Algorithm 14.4 so that it will compute X_k and Z_k for points on an elliptic curve in the Montgomery form.

14.5 Assume that if 1801 is prime, then the order of $E(-1, 0)/1801$ is 1872. Prove that 1801 is prime by the elliptic curve method.

14.6 Prove that if n is odd and equal to $x^2 + y^2$ for integers x and y, then it is congruent to 1 modulo 4. Prove that if n is not divisible by 3 and it is equal to $x^2 + xy + y^2$ for integers x and y, then it is congruent to 1 modulo 3.

14.7 Find the reduced quadratic form equivalent to each of the following:

$$3x^2 - 5xy + 7y^2, \quad 37x^2 + 39xy + 11y^2, \quad 245x^2 + 689xy + 1013y^2.$$

14.8 Let $ax^2 + bxy + cy^2$ be any quadratic form for which $b > a > 0$. Show that there is a basic transformation which will reduce the absolute value of the coefficient of xy. Use this to prove that every positive quadratic form is equivalent to a reduced form.

14.9 Find a sequence of basic transformations to get from

$$3039x^2 + 2415xy + 481y^2$$

to

$$11635x^2 + 12873xy + 3561y^2.$$

14.10 Prove that each of the three basic transformations leaves the discriminant unchanged.

14.11 Let $f(x, y) = ax^2 + bxy + cy^2$ be a reduced positive quadratic form. Show that a and c are the smallest positive integer values of n for which $f(x, y) = n$ has a solution. (*Hint:* Show that if $x \geq -y \geq 1$ and $f(x, y) \neq a$ or c, then $f(x - 1, y) < f(x, y)$.)

14.12 Use the result of Exercise 14.11 to prove that every positive quadratic form is equivalent to a unique reduced quadratic form.

14.13 Find all reduced quadratic forms of discriminant -39.

14.14 Solve $x^2 + y^2 = n$ for the following prime values of n:

$$41, \quad 409, \quad 5881, \quad 12\,541, \quad 4\,332\,721.$$

14.15 Solve $x^2 + xy + y^2 = n$ for the following prime values of n:

$$43, \quad 547, \quad 5011, \quad 12\,409, \quad 1\,554\,841.$$

14.16 Factor 29 in the complex integers.

14.17 Let $a + bi$ and $c + di$ be any two complex integers. Show that we can always find two other complex integers, say $u + vi$ and $r + si$ such that

$$a + bi \;=\; (u + vi) \times (c + di) + r + si, \quad \text{and}$$
$$0 \;\le\; N(r + si) < N(c + di).$$

(*Hint:* Let $x + yi$ be any complex number and choose a complex integer $m + ni$ such that $|x - m| \le 1/2$, $|y - n| \le 1/2$. Show that the norm of $(x + yi) - (m + ni)$ is strictly less than 1.)

14.18 Describe a Euclidean algorithm for the complex integers. Use it to prove that if $f + gi = gcd(a + bi, c + di)$ (greatest common divisor means the common divisor with the largest norm), then there exist complex integers $w + xi$ and $y + zi$ such that

$$f + gi = (a + bi) \times (w + xi) + (c + di) \times (y + zi).$$

14.19 Using the result of Exercise 14.18, prove that in the complex integers factorization is unique up to order and multiplication by ± 1, $\pm i$.

14.20 Why doesn't the fact that

$$13 = (2 + 3i) \times (2 - 3i) = (3 + 2i) \times (3 - 2i)$$

contradict uniqueness of factorization in the complex integers?

14.21 Factor 163 in the cubic integers.

14.22 Prove that $a + b\omega$ is a unit in the cubic integers if and only if it has norm equal to 1 (see Exercises 1.1, 10.24, and 10.25). Use this fact to prove that ± 1, $\pm \omega$, and $\pm \omega^2$ are the only units in the cubic integers.

14.23 Let $a + b\omega$ and $c + d\omega$ be any two cubic integers. Show that we can always find two other cubic integers, say $u + v\omega$ and $r + s\omega$ such that

$$
\begin{aligned}
a + b\omega &= (u + v\omega) \times (c + d\omega) + r + s\omega \quad \text{and} \\
0 &\leq N(r - s\omega) < N(c + d\omega).
\end{aligned}
$$

(*Hint:* Let $x + y\omega$ be any cubic number and choose a cubic integer $m + n\omega$ such that $|x - m| \leq 1/2$, $|y - n| \leq 1/2$. Show that the norm of $(x + y\omega) - (m + n\omega)$ is strictly less than 1.)

14.24 Describe a Euclidean algorithm for the cubic integers. Use it to prove that if $f + g\omega = gcd(a + b\omega, c + d\omega)$ (greatest common divisor means the common divisor with the largest norm), then there exist cubic integers $m + n\omega$ and $y + z\omega$ such that

$$
f + g\omega = (a + b\omega) \times (m + n\omega) + (c + d\omega) \times (y + z\omega).
$$

14.25 Using Exercise 14.24, prove that factorization in the cubic integers is unique up to order and multiplication by units.

14.26 Why doesn't the fact that

$$
13 = (1 + 4\omega) \times (-3 - 4\omega) = (4 + 3\omega) \times (1 - 3\omega)
$$

contradict the uniqueness of factorization?

14.27 Compute the following fourth power symbols:

$$
(1 + i/7)_4, \quad (11/1 + 10i)_4, \quad (1 + 4i/2 + 3i)_4, \quad (28 + 37i/1051)_4.
$$

14.28 For each ordinary prime between 2 and 30, find a complex prime which divides it and is congruent to 1 modulo $2 + 2i$.

14.29 For each pair of complex primes found in Exercise 14.28, compute the fourth power symbols $(p/q)_4$ and $(q/p)_4$. Can you conjecture a reciprocity law for the fourth power symbol?

14.30 Use Theorem 14.12 to compute the order of $E(-2, 0)/17$.

14.31 Use Theorem 14.12 to compute the order of $E(3, 0)/4332721$.

14.32 Compute the following sixth power symbols:

$$
(1 - \omega/5)_6, \quad (17/1 + 3\omega)_6, \quad (3 + 7\omega/2 + 5\omega)_6, \quad (1361/26 + 41\omega)_6.
$$

14.33 For each ordinary prime between 3 and 30, find a cubic prime which divides it and is congruent to 2 modulo 3.

14.34 For each pair of cubic primes found in Exercise 14.33, compute the sixth power symbols $(p/q)_6$ and $(q/p)_6$. Can you conjecture a reciprocity law for the sixth power symbol?

14.35 Use Theorem 14.14 to compute the order of $E(0,5)/19$.

14.36 Use Theorem 14.14 to compute the order of $E(0,-1)/1554841$.

The Primes Below 5000

2	181	431	683	977	1277	1567	1877
3	191	433	691	983	1279	1571	1879
5	193	439	701	991	1283	1579	1889
7	197	443	709	997	1289	1583	1901
11	199	449	719	1009	1291	1597	1907
13	211	457	727	1013	1297	1601	1913
17	223	461	733	1019	1301	1607	1931
19	227	463	739	1021	1303	1609	1933
23	229	467	743	1031	1307	1613	1949
29	233	479	751	1033	1319	1619	1951
31	239	487	757	1039	1321	1621	1973
37	241	491	761	1049	1327	1627	1979
41	251	499	769	1051	1361	1637	1987
43	257	503	773	1061	1367	1657	1993
47	263	509	787	1063	1373	1663	1997
53	269	521	797	1069	1381	1667	1999
59	271	523	809	1087	1399	1669	2003
61	277	541	811	1091	1409	1693	2011
67	281	547	821	1093	1423	1697	2017
71	283	557	823	1097	1427	1699	2027
73	293	563	827	1103	1429	1709	2029
79	307	569	829	1109	1433	1721	2039
83	311	571	839	1117	1439	1723	2053
89	313	577	853	1123	1447	1733	2063
97	317	587	857	1129	1451	1741	2069
101	331	593	859	1151	1453	1747	2081
103	337	599	863	1153	1459	1753	2083
107	347	601	877	1163	1471	1759	2087
109	349	607	881	1171	1481	1777	2089
113	353	613	883	1181	1483	1783	2099
127	359	617	887	1187	1487	1787	2111
131	367	619	907	1193	1489	1789	2113
137	373	631	911	1201	1493	1801	2129
139	379	641	919	1213	1499	1811	2131
149	383	643	929	1217	1511	1823	2137
151	389	647	937	1223	1523	1831	2141
157	397	653	941	1229	1531	1847	2143
163	401	659	947	1231	1543	1861	2153
167	409	661	953	1237	1549	1867	2161
173	417	673	967	1249	1553	1871	2179
179	421	677	971	1259	1559	1873	2203

2207	2543	2857	3251	3581	3923	4273	4657
2213	2549	2861	3253	3583	3929	4283	4663
2221	2551	2879	3257	3593	3931	4289	4673
2237	2557	2887	3259	3607	3943	4297	4679
2239	2579	2897	3271	3613	3947	4327	4691
2243	2591	2903	3299	3617	3967	4337	4703
2251	2593	2909	3301	3623	3989	4339	4721
2267	2609	2917	3307	3631	4001	4349	4723
2269	2617	2927	3313	3637	4003	4357	4729
2273	2621	2939	3319	3643	4007	4363	4733
2281	2633	2953	3323	3659	4013	4373	4751
2287	2647	2957	3329	3671	4019	4391	4759
2293	2657	2963	3331	3673	4021	4397	4783
2297	2659	2969	3343	3677	4027	4409	4787
2309	2663	2971	3347	3691	4049	4421	4789
2311	2671	2999	3359	3697	4051	4423	4793
2333	2677	3001	3361	3701	4057	4441	4799
2339	2683	3011	3371	3709	4073	4447	4801
2341	2687	3019	3373	3719	4079	4451	4813
2347	2689	3023	3389	3727	4091	4457	4817
2351	2693	3037	3391	3733	4093	4463	4831
2357	2699	3041	3407	3739	4099	4481	4861
2371	2707	3049	3413	3761	4111	4483	4871
2377	2711	3061	3433	3767	4127	4493	4877
2381	2713	3067	3449	3769	4129	4507	4889
2383	2719	3079	3457	3779	4133	4513	4903
2389	2729	3083	3461	3793	4139	4517	4909
2393	2731	3089	3463	3797	4153	4519	4919
2399	2741	3109	3467	3803	4157	4523	4931
2411	2749	3119	3469	3821	4159	4547	4933
2417	2753	3121	3491	3823	4177	4549	4937
2423	2767	3137	3499	3833	4201	4561	4943
2437	2777	3163	3511	3847	4211	4567	4951
2441	2789	3167	3517	3851	4217	4583	4957
2447	2791	3169	3527	3853	4219	4591	4967
2459	2797	3181	3529	3863	4229	4597	4969
2467	2801	3187	3533	3877	4231	4603	4973
2473	2803	3191	3539	3881	4241	4621	4987
2477	2819	3203	3541	3889	4243	4637	4993
2503	2833	3209	3547	3907	4253	4639	4999
2521	2837	3217	3557	3911	4259	4643	
2531	2843	3221	3559	3917	4261	4649	
2539	2851	3229	3571	3919	4271	4651	

Index

Undergraduate Texts in Mathematics

Thorpe: Elementary Topics in Differential Geometry.

Toth: Glimpses of Algebra and Geometry. *Readings in Mathematics.*

Troutman: Variational Calculus and Optimal Control. Second edition.

Valenza: Linear Algebra: An Introduction to Abstract Mathematics.

Whyburn/Duda: Dynamic Topology.

Wilson: Much Ado About Calculus.